INTERNATIONAL CENTRE FOR MECHANICAL SCIENCES

COURSES AND LECTURES - No. 18

GIUSEPPE LONGO
UNIVERSITY OF TRIESTE

SELECTED TOPICS IN INFORMATION THEORY

LECTURES HELD AT THE DEPARTMENT
OF AUTOMATION AND INFORMATION
SEPTEMBER - OCTOBER 1969

UDINE 1973

SPRINGER-VERLAG WIEN GMBH

Originally published by Springer-Verlag Wien-New York 1972

ISBN 978-3-211-81166-5 ISBN 978-3-7091-2850-3 (eBook)

DOI 10.1007/978-3-7091-2850-3

Foreword.

What I had in mind in preparing this course on " Selected Topics in Information Theory " was first of all to give a general idea of what Information Theory is to people who are not acquainted with it, and secondly to introduce the audience to a very limited number of particular problems which could be of interest also for those who are already a little more familiar with the theory.

Although these problems are very few, nevertheless they have not been treated as deeply as they deserve, because time was limited. In fact I have preferred to sacrifice somewhat the completeness than to treat just one problem very deeply. I hope that in so doing I have made the listeners appreciate ideas and concepts of Information Theory more extensively.

I am very indebted to CISM and particularly to the Secretary General, prof. Luigi Sobrero, for giving me the opportunity of delivering this course. I dedicate the course to prof. O. Onicescu, who introduced me into the field of Information Theory.

Udine, 1969 G. Longo

Contents.

Pages

Introduction.

We shall restrict ourselves to the mathematical aspects of Information Theory, but nevertheless we shall have in mind a very precise communication model, from which the whole theory draws its full justification. Indeed it can be said that Information Theory stemmed from the practical problem of transmitting data with as great an accuracy and efficiency as possible. Moreover it is in the field of communication techniques that Information Theory has proved most clearly to be an indispensable tool. The design of the various parts of a communication link, or of the signals to be transmitted through such a link, or of the coding and decoding equipments is accomplished in the best way with the help of Information Theory.

There are many other areas where Information Theory has been recognized as having a great importance, and its concepts are being introduced in many fields which are far away from Communication Theory.

One remarkable thing is that the mathematical structure of Information Theory has brought about methods and concepts which turned out to be very useful in other branches of mathematics, for instance in Ergodic Theory or in Statistics.

This course cannot obviously cover all the vast field of Information Theory, and does not intend to. Its aim is rather to make the audience be acquainted with some particul ar problem of mathematical character, without loosing sight, when possible, of the ideas to be found behind the formulae.

A very general survey of the mathematical model of a communication link is presented in Part One, and a statistical characterization of the source, channel and receiver is given for the particularly simple case of the discrete memoryless channel. Quantities like entropy, equivocation, rate and capacity are defined and interpreted in terms of information transmission.

In Part Two the problem of computing the channel capacity is presented in some detail, first with reference to a symmetric channel and then for any discrete memoryless channel having a square nonsingular channel matrix. In the general case the problem of computing the capacity is a complicated one and must be faced using some procedures of numerical analysis, which go beyond the limits of this course.

In Part Three the entropy - already defined in Part One for a finite scheme - is considered in greater detail. Along with Shannon's entropy, the entropies of positive order introduced by Rényi are considered, as well as the informational energy defined by Onicescu. Both the discrete and the continuous case are considered.

In Part Four a formal description of dicrete sources and channels is presented along with the concept of "typical sequences" and with the source - coding theorem of Shannon.

Part One

General Survey of the Mathematical Model for a Communication Channel.

1. 1. Introduction.

Information Theory is concerned with the generation, transmission and processing of certain quantities which are assumed to represent in some way an "amount of information". Of course a definition of "information" in general terms is by no means obvious, but since Information Theory has developed mostly in a precise though restricted direction, it is always possible to make reference to a well-determined meaning of "information". Such a meaning, originated from the pioneering work of C.E. Shannon [1] , is primarily concerned with the transmission problems, and has a statistical, or probabilistic character, which makes that modern branch of Communication Theory which embodies Information Theory be called Statistical Communication Theory [2] .

Information Theory deals with the mathematical model of communication systems and with the mathematical description of the quantities by means of which the efficiency of those systems is measured. Terms like "source", "channel", "entropy", "capacity", etc. are very common in Information Theory, and after a general introduction we shall focus on some of these con cepts to provide some insight in their nature.

1. 2. The Model for the Communication Link.

The following model is generally adopted in Communication Theory for the illustration of a communication system or link.

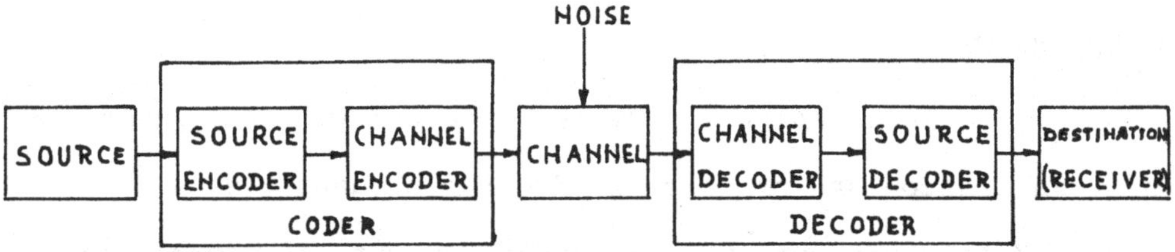

Here "source" is the device (or person) that generates the information to be trasmitted; "channel" is the physical device (a waveguide, a coaxial cable, the free space, etc.) which permits the transmission; "destination" is the device (or person) that receives the information for its purposes*.

* Remark that in practice the attribution of the equipments of a real communication channel to the sources or to the channel (and to the channel or to the receiver) is by no means uniquely determined. In other words it is largely a matter of convenience (and perhaps of opinion) where the source stops and the channel begins.

It is easily realized that generally the information cannot be sent through the channel in the form the source generates it. Information must be suitably "coded", to receive a form which can be transmitted through the channel. The channel output, in turn, must be "decoded", i.e. transformed into a form suitable for the receiver*.

The blocks marked "coder" and "decoder" are divided into two distinct subblocks, in order to make clear the distinction between the coding and decoding operations pertinent to the source-destination pair and those pertinent to the channel.

---

* To be more precise, two operations are to be performed in the coder (and respectively in the decoder); there must be a kind of transformation in the physical nature of the signals emitted by the source (e.g. if the source is the speaker, his voice cannot be broadcast directly, but must be in some way transformed into an electrical signal); and then there must be the very "coding", which operates on the signals after their transformation (e.g. if we want to transmit some continuously varying signal we can sample it with a frequency depending on its bandwidth and - here begins coding - transmit just the samples, which are numbers, in binary form up to the required accuracy: we substitute to the samples their binary representation).

The source encoder transforms the output of the source (sequence of messages) in a sequence of binary digits, which are in turn transformed by the source decoder into a hopefully acceptable reproduction of the source output and made available to the receiver.

Thus both the input of the channel encoder and the output of the channel decoder are sequences of binary digits, and this regardless of the characteristics of the pair source--destination. Of course this fact is of great importance, since it permits to design the coding and decoding equipments for the channel independently of those for the source-destination pair. See [3] .

1. 3. Accuracy Requirements.

What is the purpose of a communication link? Generally speaking it must serve to reproduce the output of the source in a place convenient to the user, and from this point of view the channel must be one of the elements of our communication chain. But when we say "to reproduce the source output", we must also add that obviously a perfect reproduction is never required. On the contrary, we expect a reproduction which meets certain specific criteria. These criteria are very different from case to case, according to the purposes for which the transmission is made. This point is perhaps clarified if we think of machine-to-machine communication: here the accuracy of the transmission must match with the precision which is required by the receiver in order to process the information.

The existence of this kind of "fidelity criterion" has a very important consequence [3] : in fact let us think of all the possible outputs of the source in a given time interval (these outputs may well constitute a continuous set) and introduce the following equivalence relation: two outputs x and y are said to be equivalent when the given fidelity criterion allows no distinction between $f(x)$ and $f(y)$ ($f(z)$ is the input at the destination when z has been transmitted).

The equivalence classes with respect to this equivalence relation may be represented each by one, suitably chosen, of its members and the transmission process may well be performed by means of this "quotient set" without loosing anything with respect to the previous way of transmitting. Once the equivalence classes are defined with respect to the given fidelity criterion, all that the source encoder is expected to do, is to indicate to the source decoder (through the channel) the class to which the source output belongs. Once the source decoder knows this class, it generates a signal representative of the class itself, and conveys it to the receiver. Clearly this operation is repeated for any output of the source.

In so doing we are always able to reduce the set of possible "distinguishable" source outputs to a discrete set, and this set is called the set of the source "messages".

1. 4. Encoding for Minimum Average Length. Source Rate.

What is the task of the source encoder? Once the source has delivered its output, the endoder must assign it to an equivalence class and then emit a binary sequence which determines uniquely that class. So we must have as many different binary sequences as there are classes, and these sequences can be all of the same length (uniform coding) or of different length (nonuniform coding). In the latter case no sequence can be the initial part of a longer sequence, and a minimum average length of the sequences fed into the channel can be achieved when the shorthest sequences are associated to the messages (or equivalence classes) which are most likely. More precisely the minimum average length (minimum average number of binary digits per message) is obtained when the length of each word (binary sequence) is - approximately - equal to the lorarithm to the base 2 of the reciprocal of the message probability.

It is rather obvious that the average length of the code words depends on the message probabilities and therefore on the number of source outputs contained in each equivalence class (message). These probabilities depend on the statistical properties of the source and on the way we group the source outputs into classes (for a given fidelity criterion there is in general more than one way of doing this groupage).

This circumstance allows us to minimize (at least in principle) the average number of output binary digits.

Equivalently, if we assume that the source delivers the messages with a certain rate (say m messages per second), we may try to minimize the average number of output binary digits per second.

This least average number of binary digits per second which have to be transmitted in order to reproduce the source output according to the given fidelity criterion is called the rate R at which the source generates the information.

We remark that the rate R is uniquely specified by the source (i.e. by its statistics and output set) and by the fidelity criterion, once the number of source outputs per second is specified.

1. 5. The Noisy Channel. Reduction of Transmission Errors.

Now we give an outline of the box marked "channel". This channel connects the source with the user, through suitable coding and decoding apparatuses, and may be thought of as a device able to generate any event of a given class and to observe it. Any physical channel involves the presence of energy, since energy is present in any kind of physical observation, and therefore the presence of a medium through which energy can propagate. In other words, the transmission of information always requires propagation of energy; and conversely any propagating energy can be used to transmit information [4, 5, 6] .

It is well known that any kind of physical observation is subject to measurement uncertainties and errors, resulting from random disturbances. These random disturbances (or noise) have the effect that observations of different events may lead to the same result, so making uncertain the inference as to which event actually occurred.

If the channel is fed with binary digits, say 0 and 1, the noise makes it possible that sometimes a 0 is received as a 1 and conversely. When this happens, we speak of an error in the transmission, since when we receive a 0 we conclude that a 0 was transmitted, independently of the digit actually transmitted. Since the noise has random characteristics, there is a certain probability $P(e)$ that any particular binary digit is decoded incorrectly.

So far we have seen that there are two parameters that characterize the transmission process:

1) the transmission rate R (number of binary digits per second input into the channel encoder);
2) the reception accuracy, as measured by the probability of error per digit $P(e)$.

If the channel is noisy, we may always think of $P(e)$ as being $< 1/2$. Now suppose $R = 1$ and $P(e) = 1/4$; if we feel that 1/4 is a too high probability of incorrect decoding, we can lower it as follows: we repeat the transmission of each digit an odd number of times, say 3 , and decode any received 3-digit sequence as 1 (or 0) if two or three digits in it are 1's (or 0's).

In this way, we have an error only if two or three digits are received incorrectly. If the noise is uncorrelated and the channel is memoryless and time-invarying, the probability of having two or three errors in three digits is:

$$\binom{3}{2}\frac{3}{4}\cdot\left(\frac{1}{4}\right)^2+\left(\frac{1}{4}\right)^3=\frac{10}{64} \quad ,$$

which is less than 1/4.

The price we must pay for this reduction in the error probability is that if we cannot transmit more than one digit/sec through the channel, now it takes 3 seconds to transmit one digit output by the source; if we want to synchronize the source with the channel, we must slow down the rate of the source to 1/3 of its previous value.

Of course this procedure can be generalized to an arbitrary odd number 2n+1 of repetitions of the same digit, so making $P(e)$ go to zero when n goes to ∞ .

Of course we have:

Prob (error in n+1 or more digits) =

$$= \sum_{n+1}^{2n+1}{}_k \binom{2n+1}{k} \alpha^k (1-\alpha)^{2n+1-k} \quad ,$$

where $\alpha = P(e)$ is the error probability for a single digit.

Now let E_{2n+1} be the number of erroneous digits in the (2n+1)-digit sequence; by the weak law of large numbers, the ratio

$$\frac{E_{2n+1}}{2n+1} \ .$$

converges in probability to α when $n \to \infty$. On the other hand the expected value of the number of erroneous digits is α (2n+1) $<$ $<$ n+1, so that $\frac{n+1}{2n+1} = \alpha + \varepsilon$.

Thus:

$$\text{Prob (error in n+1 or more digits)} = \text{Prob} (E_{2n+1} \geq n+1) =$$

$$= \text{Prob} \left(\frac{E_{2n+1}}{2n+1} \geq \frac{n+1}{2n+1} \right) = \text{Prob} \left(\frac{E_{2n+1}}{2n+1} \geq \alpha + \varepsilon \right) \to 0$$

as $n \to \infty$.

Apparently, while making arbitrarily small the error probability, we also reduce the effective transmission rate to zero. This of course is not very pleasant, but, according to our intuition, it is the only possibility. Well, at this point it is perhaps surprising that the fundamental theorem of Information Theory states essentially that it is not necessary to reduce the transmission rate to zero, in order to achieve arbitrarily reliable transmission, but only to a positive number called the channel capacity. Of course, in order to obtain in practice this fine result we must choose rather complicated ways of associating the binary sequences to the source messages. This association, as we have said, is called "coding", and coding theory has developed into a very extensive field of its own.

Of course also the afore-described way of associating to each 1 (or 0) a sequence of 2n+1 1's (0's) is a form of coding, although very inefficient.*

---

* We wish to point out that in the case of noiseless channel the

One essential aspect of coding is that the messages of the sources are no longer encoded one by one, but rather in blocks of m of them, and it is just this procedure that, in the limit for $m \to \infty$, permits arbitrarily high reliability in the transmission with positive rate.

We remark that the coding theorem is an "existence" theorem, and does not give any recipe for actually constructing the codes; moreover no general procedure for constructing codes has so far been found.

Even from the few words we have spent about the coding theorem one can perhaps get the impression that Information Theory is asymptotic in its nature, at least as far as concept like source, channel, error probability, coding etc. are concerned. Indeed asymptotic methods have a great part in the theory, and most results which are concerned with the applications are based on asymptotic results.

---

* purpose of coding was to minimize the average length of the code words (i.e. to improve the channel efficiency in terms of time), while in the case of noisy channel the purpose of coding is to lower the error probability in reception, even disregarding at a certain extent the channel efficiency (provided this efficiency is not reduced to zero).

1. 6. Entropy and Information. The Finite Case.

Now we turn to a more precise definition of such concepts as "rate" of a source and "capacity" of a channel. This task is accomplished perhaps in the best way by defining entropy and information from a formal point of view. The different definitions, properties and propositions that we shall state, will be illustrated when possible in terms of information transmission, channels, sources and the like.

We must point out preliminarily that the transmission of information is essentially a statistical process, i.e. both the source and the channel work on a statistical basis. So the source is, from a mathematical point of view, a random process, which in the simplest cases reduces to a "finite scheme", with successive trials that we may consider as equispaced in time. Sources of the Markov type have received considerable attention. In its turn the channel may be considered as a family of conditional probability densities, one for each possible input, which rule the generation of the outputs. In the simplest cases both the input and the output space are discrete or even finite, so that the conditional probabilities may be arranged in a rectangular matrix, the "channel matrix".

In defining entropy and information we shall restrict ourselves for the moment to the finite case.

Consider a finite probability space and let $x_1, x_2, \ldots, x_n$ be its (mutually exclusive) possible simple events. (Equivalently consider an experience whose possible outcomes are $x_1, x_2, \ldots\ldots x_n$). Let $p_1, p_2, \ldots\ldots, p_n$ be the probabilities of the events $x_1, x_2, \ldots\ldots, x_n$ respectively. We have then to do with a finite scheme X:

$$X = \begin{pmatrix} x_1 & x_2 & \ldots & x_n \\ p_1 & p_2 & \ldots & p_n \end{pmatrix}, \quad p_i \geq 0 \ , \quad \sum_1^n{}_i \, p_i = 1 \ .$$

In case the probability distribution is not degenerate (i.e. $p_i = \delta_{ij}$)* the random variable ($\mathcal{R}V$) X associated with the scheme X has an amount of uncertainty, which proves to be suitably expressed by the number:

$$H(X) = H(p_1, p_2, \ldots\ldots, p_n) = -\sum_1^n{}_i \, p_i \log p_i \ , \qquad (1.1)$$

where, as customary in Information Theory, the base of the logarithms is 2 . The number $H(X)$, considered as a function of the p_i and defined in the domain $S = (p_1, \ldots, p_n : p_i \geq 0, \sum_1^n{}_i \, p_i = 1)$ is called the <u>entropy</u> of the finite scheme X or of the corresponding $\mathcal{R}V$ X .

<u>Remark 1.</u> Observe that $H(X)$ is not a function of the actual values $x_1, \ldots\ldots, x_n$ taken on by the $\mathcal{R}V$ X , but only of the probability distribution. Actually if x and y are two distinct $\mathcal{R}V$'s having the same probability distribution, then

---

* We shall make use of the δ_{xs} symbol of Kronecker; here i is a current index and j is fixed.

$H(x) = H(y)$ matter what the possible values of x and y are.

Remark 2. Observe that if some among the p's is zero, then we let p log p = 0 (according to the continuity definition), and therefore the corresponding terms in the summation of (1.1) are deleted.

The entropy (1.1) has some desirable properties which we should require from an "uncertainty function". First of all, it is apparent that:

$$H(X) \geq 0 \quad , \tag{1.2}$$

and the equality sign holds if and only if (iff) $p_i = \delta_{ij}$. This of course matches with the fact that in this case there is no uncertainty at all.

We may think of a source whose possible outputs are $x_1, x_2, \ldots\ldots, x_n$, that are generated independently of each other according to the probability distribution p_1, p_2, ..., p_n. $H(X)$ is then the average uncertainty as to which symbol is emitted by the source, or also the average information provided by the source itself. Then this quantity cannot be negative, as expressed by eq. (1.2).

Another remarkable property of $H(X)$ is the following:

$$H(X) \leq H\left(\frac{1}{n}, \frac{1}{n}, \ldots, \frac{1}{n}\right) = \log n \quad , \tag{1.3}$$

where the equality holds iff $p_i = \frac{1}{n}$ $(i = 1,2,\ldots,n)$, i.e. for the uniform distribution*. This is again very sensible, since the uniform distribution corresponds to the state of greatest uncertainty as to the outcome.

In the source example, when the distribution is uniform every symbol emitted provides the same information, log n, which is the greatest possible value of the information the source may provide when the successive letters are independent of each other.

Remark. From (1.1) it is apparent that $H(X)$ is an expected value, precisely of the function $\log \frac{1}{p_i}$. It is known [3] that this logarithm of the inverse of p_i is the self-information $I(x_i)$ of the event x_i having probability p_i. So we may say that $H(X)$ is the expected value of the self-information in the finite scheme X. This name of self-information has its justification in the fact that the quantity

$$I(x_i ; y_k) = \log \frac{\text{Prob}(x_i / y_k)}{\text{Prob}(x_i)}$$

---

* The proof follows from the inequality $\varphi\left(\frac{1}{n}\sum_1^n{}_i \alpha_i\right) \leq \frac{1}{n}\sum_1^n{}_i \varphi(\alpha_i)$, which holds for any concave function φ, and in particular for the function $\varphi(x) = x \log x$. Then replacing the a_i by the p_i:

$$\left(\frac{1}{n}\sum_1^n{}_i p_i\right) \log \left(\frac{1}{n}\sum_1^n{}_i p_i\right) = \frac{1}{n} \log \frac{1}{n} \leq \frac{1}{n}\sum_1^n{}_i p_i \log p_i = -\frac{1}{n} H(p_1, \ldots p_n) .$$

is called the mutual information of x_i and y_k *, i.e. $I(x_i ; y_k)$ is the amount of information that the outcome of y_k provides about the outcome of x_i. When y_k determines uniquely x_i, so that Prob $(x_i/y_k)=1$, then the mutual information $I(x_i ; y_k)$ reduces to the self-information $I(x_i)$. This self-information is apparently the greatest amount of information that any event y_k may provide about x_i (in fact Prob (x_i/y_k) $\leq$ 1 implies $I(x_i ; y_k) \leq I(x_i)$), and at the same time the greatest amount of information that x_i may provide about any other y_k (in fact Prob (y_k/x_i) $\leq$ 1 implies $I(y_k ; x_i) = I(x_i ; y_k) \leq I(x_i)$).

To sum up, the entropy $H(X)$ is the average self-information of the events x_i (or of the symbols x_i, if we consider the source: when the source emits a sequence of sym bols, under certain conditions, then the self-information provided in the mean by one symbol is $H(X)$).

Now we consider besides X another finite scheme:

$$Y = \begin{pmatrix} y_1 & y_2 & \dots & y_m \\ q_1 & q_2 & \dots & q_m \end{pmatrix} \tag{1.4}$$

and the πv y associated with it, and we make the following posi tions:

Prob (x_i/y_k) = $p_{i/k}$; Prob (y_k/x_i) = $q_{k/i}$; Prob (x_i , y_k) = π_{ik}.

---

* Remark that, since $\frac{\text{Prob}(x_i/y_k)}{\text{Prob}(x_i)} = \frac{\text{Prob}(y_k/x_i)}{\text{Prob}(y_k)}$, it follows the symmetry property $I(x_i ; y_k) = I(y_k ; x_i)$.

Obviously we have:

$$\sum_1^n {}_i\, p_{i/k} = 1 \;\forall k\;; \quad \sum_1^m {}_k\, q_{k/i} = 1 \;\forall i\;; \quad \sum_1^n {}_i\, \pi_{ik} = q_k \;\forall k\;;$$
$$\sum_1^m {}_k\, \pi_{ik} = p_i \;\forall i\;; \quad \pi_{ik} = p_i\, q_{k/i} = q_k\, p_{i/k} \;. \tag{1.5}$$

Observe that the $n\,m$ numbers π_{ik} are relative to the "product" finite scheme $X \otimes Y$, whose events are

$x_1 y_1$, $x_1 y_2$, ..., $x_n y_m$.

We can apply the definition (1.1) to this product finite scheme $X \otimes Y$, obtaining:

$$H(x, y) = -\sum_1^n {}_i \sum_1^m {}_k\, \pi_{ik} \log \pi_{ik} \;. \tag{1.6}$$

Since, on the other hand, $\sum_1^n {}_i\, p_{i/k} = 1$ for every k, $p_{1/k}$, $p_{2/k}$, ... , $p_{n/k}$ is easily seen to be a probability distribution, whose entropy is given by:

$$H(x/y = y_k) = -\sum_1^n {}_i\, p_{i/k} \log p_{i/k} \tag{1.7}$$

and is called the conditional entropy of the πv x subject to the condition that the πv y takes on the value y_k . Of course, since y takes on the value y_k with probability q_k , we may consider a πv z which takes on the value $H(x/y = y_k)$ with probability q_k . The mean value of z is then:

$$\overline{H(x/y = y_k)} = H(x/y) = \sum_1^m {}_k\; q_k\, H(x/y_k) = -\sum_1^n {}_i \sum_1^m {}_k\, \pi_{ik} \log p_{i/k}\,, \tag{1.8}$$

whence:

$$H(x/y) = H(x,y) - H(y) \quad . \qquad (1.9)$$

In a quite similar way we obtain also:

$$H(y/x) = H(x,y) - H(x) \quad . \qquad (1.10)$$

Eqs. (1.9) and (1.10) may also be put into the form:

$$H(x,y) = H(x) + H(y/x) \quad , \qquad (1.11)$$

$$H(x,y) = H(y) + H(x/y) \quad , \qquad (1.12)$$

whence:

$$H(x) + H(y/x) = H(y) + H(x/y) \quad , \qquad (1.13)$$

$$H(x) - H(x/y) = H(y) - H(y/x) \quad . \qquad (1.13')$$

Consider now the case when the πV'sx and y are independent, i.e. $\pi_{ik} = p_i q_k$, or equivalently $p_{i/k} = p_i$ and $q_{k/i} = q_k$ Then from eq. (1.7) it follows that $H(x/y = y_k) = H(x)$ for every k, and consequently from eq. (1.8) we have:

$$H(x/y) = H(x) \quad \text{(case of independence)} \qquad (1.14)$$

and from eq. (1.12):

$$H(x,y) = H(x) + H(y) \qquad \text{(case of independance)} \qquad (1.15)$$

Of course, under the same assumptions we obtain also $H(y/x) = H(y)$, whence eq. (1.15) again.

Since $H(x/y)$ is the mean value of the non-negative quantity (1.7), we have:

$$H(x/y) \geq 0 \quad , \qquad (1.16)$$

and the equality holds iff for every k there is a unique i_k in (1.7) for which $p_{i_k/k} = 1$, while $p_{i/k} = 0$ for $i \neq i_k$ (i.e. $p_{i/k} = \delta_{i\, i_k}$).

Moreover the inequality:

$$H(x/y) \leq H(x) \quad . \qquad (1.17)$$

holds, with equality iff x and y are independent (cf. (1.14)).

Similarly of course:

$$0 \leq H(x/y) \leq H(y) \quad . \qquad (1.17')$$

The only thing to prove is (1.17). Consider the difference

$$H(X/Y) - H(X) = -\sum_1^n{}_i \sum_1^m{}_k q_k p_{i/k} \log p_{i/k} + \sum_1^n{}_i p_i \log p_i =$$
$$= \sum_1^n{}_i \sum_1^m{}_k \pi_{ik} \log \frac{p_i}{p_{i/k}} \leq \sum_1^n{}_i \sum_1^m{}_k \pi_{ik} \left(\frac{p_i}{p_{i/k}} - 1\right) \log e =$$
$$= \left(\sum_1^n{}_i \sum_1^m{}_k q_k p_i - \sum_1^n{}_i \sum_1^m{}_k \pi_{ik}\right) \log e = 0 \quad ,$$

where the inequality

$$\log_e x \leq x - 1 \tag{1.18}$$

has been utilized.

Since in eq. (1.18) the equality holds iff $x = 1$ eq. (1.17) holds, with equality iff $p_{i/k} = p_i$ for every i and k .

Now we try to give an interpretation of equations (1.6) to (1.17') in terms of information transmission. Suppose our source is able to generate M different messages, which form a set $A = (a_1, a_2, \ldots\ldots, a_M)$ to which the probabilities $\text{Prob}(a_k) = p_k$ are associated. The coder associates to each message a_k a code word w_k (having of course the same probability p_k as a_k), consisting of a sequence of symbols (letters) belonging to a given alphabet. Let B be the number of letters in the alphabet, and n_k the length of the k-th code word, w_k .

Next define the average length of the message as:

$$\bar{n} = \sum_1^M{}_k p_k n_k \quad . \tag{1.19}$$

The source may be thought of as a finite scheme:

$$\begin{pmatrix} a_1 & a_2 & \ldots & a_M \\ p_1 & p_2 & \ldots & p_M \end{pmatrix} \tag{1.20}$$

whose entropy $H(A)$ represents the average amount of information that we must provide about one message to specify it uniquely. Now inequality (1.3) reads:

$$H(A) \leq \log M \quad , \tag{1.21}$$

where the equality sign holds iff the a_k have the same probability $1/M$, which is not true in general. On the other hand the symbols of the prescribed alphabet provide a maximum average information when they are used in such a way that they turn out to be equiprobable, and this maximum average value is

$$\log B \quad , \tag{1.22}$$

which we may call the "capacity of the coding alphabet".

Now with reference to eq.(1.17), we may think of $x, y \ldots$ as of the successive letters (symbols) of a code word, and eq. (1.17) tells us that the average amount of information provided by x (i.e. by one symbol) cannot be increased by making x depend statistically on the identity of (the preceding symbol) y . Eq. (1.17) may be easily generalized:

$$H(z/x, y) \leq H(z/y) \leq H(z) \quad , \tag{1.23}$$

where $x, y, z,$ are three successive symbols of a code word.

We can conclude that:

$$\bar{n} \log B \geq H(A) \quad . \tag{1.24}$$

which means that if we are able to encode all the messages $a_1, \ldots, a_M$ of the source, then the average information of a message, $H(A)$, cannot exceed the mean length $\bar{n}$ times the maximum average information of one code symbol. Eq.(1.24) provides a lower bound on $\bar{n}$:

$$\bar{n} \geq \frac{H(A)}{\log B} \quad , \tag{1.25}$$

which reduces to

$$\bar{n} \geq H(A) \tag{1.26}$$

in the case of a binary code (B = 2).

The rules to be followed in obtaining values of $\bar{n}$ as closed as possible to the bound in (1.25) or in (1.26) are two: first of all every code symbol should appear in every position with the same probability, and then every code symbol should not depend statistically on the previous one. Remark that we do not require - and we could not indeed - that the messages of the source are equiprobable, since the source is given once for all. Some techniques [3,7] have been proposed to reduce $\bar{n}$ to its lower bound (1.25), which is however possible only in par ticular cases. The fundamental idea involved is that of associating the shortest code words with the most likely messages, taking into account the two afore-mentioned rules.

We should however point out that if the code symbols are transmitted with (approximately) the same probability and independently, one error in the transmission (due to the presence of noise, which can transform one symbol into another one) cannot be detected and consequently not corrected, which is highly undesirable. For this reason all the techniques which deal with the reduction of $\bar{n}$ to its lower bound $H(A)/\log B$ are known as the "noiseless coding techniques", since when the channel is noiseless the aim is apparently that of transmitting the information generated by the source in the fastest way.

1. 7. Channel Equivocation and Channel Capacity.

Now suppose the channel is noisy, i.e. for each symbol x_i input to the channel, there is not a unique output symbol y_i which can be received.

Formally we may say that a discrete channel (without memory) is a pair of finite abstract spaces:

$$X = (x_1, x_2, \ldots x_n) \; ; \qquad Y = (y_1, y_2 \ldots, y_n) \; ;$$

together with a family $q_{k/i}$ = Prob y_k / x_i $i = (1, \ldots n$; $k = 1, \ldots, m$) of conditional probability distributions over Y , one for each x_i .* If we add a probability distribution to X, we get a source, which we call again X :

$$X = \begin{pmatrix} x_1 \, x_2 \, . \, . \, . \, x_n \\ p_1 \, p_2 \, . \, . \, . \, p_n \end{pmatrix} \quad .$$

Taking into account the eq.s in (1.5), we have

$p_{i/k}$ = Prob (x_i was transmitted when y_k is received) =

$$\frac{\pi_{ik}}{q_k} = \frac{\pi_{ik}}{\sum_1^n{}_i \pi_{ik}} = \frac{p_i \, q_{k/i}}{\sum_1^n{}_i (p_i \, q_{k/i})} \quad .$$

Now suppose a particular y_k , say $y_{\bar{k}}$, is received; all our knowledge as to which x_i was actually transmitted is contained in the conditional probability distribution $p_{i/\bar{k}}$, and the mutual information $-\log p_{i/\bar{k}}$ is the amount of information provided by $y_{\bar{k}}$ about each x_i . The average mutual information $-\sum_1^n{}_i \, p_{i/\bar{k}} \log p_{i/\bar{k}} = H(x/y = y_{\bar{k}})$ is the amount of information we need on the average to determine the transmitted x_i when $y_{\bar{k}}$ is received; and averaging this average over all the y_k , we obtain, according to eq.(1.9):

$$H(x/y) = -\sum_1^m{}_k \, q_k \sum_1^n{}_i \, p_{i/k} \log p_{i/k} = -\sum_1^n{}_i \sum_1^m{}_k \pi_{ik} \log p_{i/k} \quad , \qquad (1.9)$$

---

* Of course the channel is noiseless if $q_{k/i}$ is either 0 or 1 for every pair (x_i , y_k), with the condition $\sum_1^m{}_k \, q_{k/i} = 1$; otherwise the channel is noisy, which does not mean that it necessarily looses information.

which is usually called the "equivocation" [8] of the channel with respect to the given probability distribution $\{p_i\}$ of the source.

In other words, every symbol x_i is completely known at the channel input (= source output), but it is known only with an uncertainty given by (1.9) at the channel output (= receiver input); and this increase in uncertainty from zero to the value given by eq. (1.9) is called the equivocation of (or introduced by) the channel, and coincides with the average amount of information lost in the channel ("average" means "per symbol").

After this, it is clear that the quantity

$$H(x) - H(x/y)$$

is the average (i.e. per symbol) amount of information received through the channel (in fact it is the average amount of information required to specify a symbol before transmission minus the average amount of information required to specify a symbol after transmission).

From eq.s (1.16) and (1.17) we get at once:

$$0 \leq H(x) - H(x/y) \leq H(x) \quad , \qquad (1.28)$$

where the equality holds on the right when any y_k uniquely specifies a x_i * i.e. when the residual average uncertainty $H(x/y)$

---

* This case corresponds to a "lossless channel"[9], which is

after transmission is zero (and therefore the residual uncertainty $H(x/y=y_k)$ is zero for every k); the equality holds on the left in (1.28) when $H(X/Y) = H(X)$, i.e. when there is no statistical dependence between the symbol received and the symbol transmitted. In the first case we have no information loss, in the second case we have no information transmission.

These remarks make it plausible to call the quantity

$$I(x;y) = H(x) - H(x/y) \tag{1.29}$$

---

* not noiseless in general, for more than one y_k may specify uniquely an x_i (the noiseless channel provides a one-one correspondence between the set of transmitted symbols and the set of the received symbols). In other words, a lossless channel is one for which the output space Y may be divided into n non-overlapping sets A_i (i = 1,......,n), whose union is the whole of Y , such that Prob (A_i/x_i) = 1. This results in saying that any output y_k uniquely determines the transmitted x_i , so that $H(x/y) = 0$. If in addition every input x_i uniquely determines the output y_k , i.e. if $H(y/x) = 0$, then every set A_i consists of only one element, and the channel is noiseless. A channel which meets the condition $H(y/x) = 0$ (i.e. Prob (y_k/x_i) is zero or one for each k and i) is said to be deterministic. Therefore a noiseless channel is a lossless and deterministic channel.

the (average) amount of information transmitted through the channel, or also the transmission rate of the channel. Remark that the quantity in (1.29) is symmetric with respect to x and y , as it follows from eq. (1.13').

The quantity (1.29), i.e. the information processed by the channel, depends on the input probability distribution $\{p_i\}$. So we may think of varying the input distribution until $I(x;y)$ reaches a maximum; this maximum of the transmission rate is called the <u>channel capacity</u> C :

$$C = \max_{\{p_i\}} \{H(x) - H(x/y)\} \qquad (1.30)$$

C is a true maximum and not a least upper bound, i.e. there is at least one input probability distribution which achieves the channel capacity. In fact the conditions $p_i \geq 0$ and $\sum_1^n {}_i\, p_i = 1$ determine a bounded closed set S in the n-dimensional Euclidean space; on the other hand $I(x;y)$ is a continuous function of the n variables p_i , and therefore must reach a maximum (as well as a minimum) value in S .

Part Two

Calculation of Channel Capacity.

2. 1. Case of Symmetric Channel.

The calculation of the capacity of a given channel is a rather difficult task in general, and we shall limit the discussion to some special cases.

Consider a symmetric channel [3] , [9] , i.e. a channel such that the columns of the channel matrix $\|q_{k/i}\|$ are (not necessarily distinct) permutations of the same set of numbers, and similarly the rows of the channel matrix are (not necessarily distinct) permutations of the same set of n numbers, with the additional condition that $\sum_{1}^{m} {}_k\, q_{k/i} = 1 \ \forall i$. If we arrange the numbers $q_{k/i}$ in a matrix as follows:

$$\begin{bmatrix} q_{1/1} & q_{2/1} \cdots & q_{m/1} \\ q_{1/2} & q_{2/2} \cdots & q_{m/2} \\ \cdots & & \\ q_{1/n} & q_{2/n} \cdots & q_{m/n} \end{bmatrix} \qquad (2.1)$$

that condition means that the sum of the elements of any row is 1, i.e. the matrix is stochastic.

The well-known binary symmetric channel is apparently a symmetric channel; in fact its matrix and its schematic representation are as follows:

$$\begin{bmatrix} 1-p & p \\ p & 1-p \end{bmatrix}$$

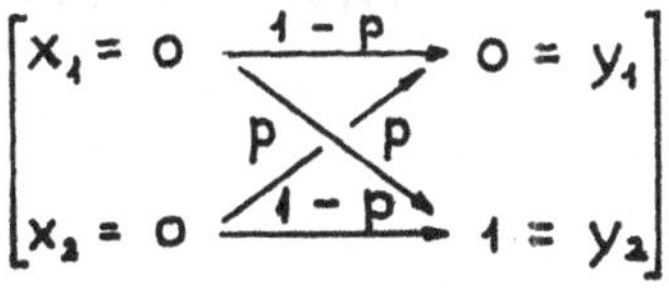

(2.2)

From the definition of symmetric channel it follows at once that for such a channel $H(y/x)$ is independent of the input distribution $\{p_i\}$. In fact, for the sake of simplicity let $(a_1, a_2, \ldots . a_n)$ be the numbers which make up the columns and $(b_1, b_2, \ldots . b_m)$ the numbers making up the rows of the matrix in (2.1). Suppose x_i has been transmitted, thus:

$$H(y/x = x_i) = -\sum_1^m {}_k b_k \log b_k \qquad (i = 1, \ldots, n) \quad ,$$

which does not depend on i, and therefore:

$$H(y/x) = \sum_1^n {}_i p_i H(y/x = x_i) = -\sum_{k=1}^m b_k \log b_k \; ; \qquad (2.3)$$

it is thus clear that in $H(y/x)$ the input distribution $\{p_i\}$ plays no role.

Now, according to (1.30), in order to find the capacity C we want to maximize the rate $H(x) - H(x/y)$, which in force of eq. (1.13') coincides with the expression

$$H(y) - H(y/x) \quad . \qquad (2.4)$$

In order to maximize the quantity in (2.4) with respect to the input distribution (for a symmetric channel), we should maximize $H(y)$, since $H(y/x)$ is constant, as we have seen it in eq. (2.3). The theorem expressed by eq. (1.3) tells us that

$$H(y) \le \log m \quad ,$$

and the maximum log m is attained when all the output symbols are equiprobable. This uniform output distribution is obtained when the input distribution is uniform; in other words:
if the input symbols of a symmetric channel are equiprobable, so are the output symbols.

To prove this statement, let $p_i = 1/n$ $(i = 1,..,n)$; then according to eq.s (1.5) we have:

$$\text{Prob}(y_k) = q_k = \sum_1^n {}_i\, \pi_{ik} = \sum_1^n {}_i\, p_i q_{k/i} = 1/n \sum_1^n {}_i\, q_{k/i} = 1/n \sum_1^n {}_i\, a_i = \text{const} .$$

The result is that if $p_i = 1/n$, then $q_k = 1/m$, and the maximum for $H(y)$ is attained.

This permits us to write the expression for the capacity of a symmetric channel in the following way:

$$C_{sym} = \log m + \sum_1^m {}_k\, b_k \log b_k \quad . \qquad (2.4)$$

In particular for the binary symmetric channel given in (2.2), the capacity is:

$$C_{BSC} = 1 - H(p, 1-p) = 1 + p \log p + (1-p) \log (1-p) \quad . \qquad (2.5)$$

2. 2. Some Properties of $I(x;y)$.

Now we turn to the calculation of the channel capacity in a more general case, and to this end we need some further properties of the function $I(X;Y)$ given by (1.29). We said that the rate $I(X;Y)$ is a function of the input distribution $\{p_i\}$. Now consider various input distributions $\{p_i^{(\ell)}\}$ and label $I_\ell(x;y)$ the rate of the channel when the source driving it operates according to the distribution $\{p_i^{(\ell)}\}$.

Then we can prove the following theorem:

Theorem: The rate $I(x;y)$ is a convex function of the input probabilities, i.e. if $t_1, t_2, \dots t_r$ are nonnegative numbers whose sum is 1 and if we let the convex linear combination of the p_i:

$$\{\bar{p}_i\} = \left\{\sum_{1}^{r}{}_\ell\, t_\ell\, p_i^{(\ell)}\right\} \qquad i = (1, \dots, n) \tag{2.6}$$

be an input distribution, then the rate $\bar{I}(X;Y)$ corresponding to the distribution $\{\bar{p}_i\}$ in (2.6) satisfies:

$$\bar{I}(x;y) \geq \sum_{1}^{r}{}_\ell\, t_\ell\, I_\ell(x;y)^{*} \tag{2.7}$$

---

* From (2.7) it follows in particular that the entropy $H(X)$ is a convex function of the input probabilities; in fact it is sufficient to consider a lossless channel, for which $H(X/Y) = 0$ i.e. $I(X;Y) = H(X)$. Since $H(X)$ does not depend on the channel but only on the source, it turns out to be convex.

To prove this convexity property of $I(x;y)$, we consider the difference between the two sides in eq. (2.7); we get:

$$\Delta I = \bar{I}(x;y) - \sum_{1}^{r}{}_{l} t_l I_l(x;y) = \bar{H}(y) - \bar{H}(y/x) - \\ - \sum_{1}^{r}{}_{l} t_l \left[H_l(y) - H_l(y/x)\right] \quad , \tag{2.8}$$

where $\bar{H}(y/x) = -\sum_1^n{}_i \sum_1^m{}_k \bar{\pi}_{ik} \log \bar{q}_{k/i}$, being $\bar{\pi}_{ik} = \sum_1^r{}_l t_l \pi_{ik}^{(l)}$ and $\bar{q}_{k/i} = \sum_1^r{}_l t_l q_{k/i}^{(l)}$. Of course $\bar{H}(y/x)$ may be written in the following form:

$$\bar{H}(y/x) = \sum_1^r{}_l t_l \left\{-\sum_1^n{}_i \sum_1^m{}_k \pi_{ik}^{(l)} \log q_{k/i}^{(l)}\right\} = \sum_1^r{}_l t_l \; H_l(y/x) \quad , \tag{2.9}$$

anf therefore (2.8) yields:

$$\Delta I = \bar{H}(y) - \sum_1^r{}_l t_l H_l(y) = -\sum_1^m{}_k \bar{q}_k \log \bar{q}_k + \sum_1^r{}_l t_l \left\{\sum_1^m{}_k q_k^{(l)} \log q_k^{(l)}\right\} = \\ = \sum_1^r{}_l t_l \left\{-\sum_1^m{}_k q_k^{(l)} \log \bar{q}_k + \sum_1^m{}_k q_k^{(l)} \log q_k^{(l)}\right\} \quad . \tag{2.10}$$

Now a well-known inequality [8], [10] states * that:

$$-\sum_1^m{}_k q_k \log q_k \leq -\sum_1^m{}_k q_k \log p_k \quad \left(q_k \geq 0, \, p_k \geq 0, \sum_1^m{}_k q_k = \sum_1^m{}_k p_k = 1\right) \tag{2.11}$$

---

* Since $\ln x \leq x - 1$ with equality iff $x = 1$, we have $\ln(q_k/p_k) \leq (p_k/q_k) - 1$ whence $q_k \ln(p_k/q_k) \leq p_k - q_k$ and summing over k from 1 to m: $\sum_1^m{}_k q_k \ln\left(\frac{p_k}{q_k}\right) \leq 1 - 1 = 0$, with equality iff $p_k = q_k$ for every k.

with equality iff $p_k = q_k$ (k = 1,....,m), and since in (2.10) the numbers t_l are nonnegative, (2.11) yields $\Delta I \geq 0$, completing the proof.

The importance of this theorem is readily seen: if we find a "stationary" point of the function $I(x;y)$ of the n variables p_i (i = 1,. . ., n) using the ordinary rules of differential calculus, we are sure that point is an absolute maximum rather than a relative maximum or a saddle point. Of course this does not entail that the absolute maximum is unique: there can be an infinity of (p_1,. . ., p_n) points achieving this maximum.

As an example consider the channel schematized in the following picture:

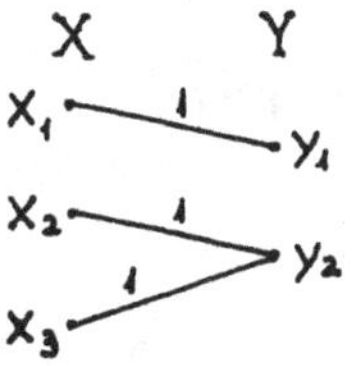

$$P(y_1/x_1) = 1 \quad P(y_2/x_2) = 1 \quad P(y_2/x_3) = 1$$

$$X = \begin{pmatrix} x_1 & x_2 & x_3 \\ p_1 & p_2 & p_3 \end{pmatrix}$$

$$\text{channel matrix} = \begin{bmatrix} 1 & 0 \\ 0 & 1 \\ 0 & 1 \end{bmatrix}$$

Using the relations in (1.5), it is not difficult to see that $I(x;y) = H(x) - H(x/y) = H(y) - H(y/x)$ is equal* to:

---

* In fact, since Prob (y_k/x_i) is 1 or 0 for each k and i, the channel is deterministic and $I(x;y)$ reduces to $H(y)$, which is easily seen to coincide with (2.12).

$$I(x;y) = -p_1 \log p_1 - (p_2+p_3)\log(p_2+p_3) \quad . \qquad (2.12)$$

Now whenever $p_1 = p_2 + p_3 = 1/2$, the expression in (2.12) has a maximum value, $\log 2 = 1$, which is the capacity of the channel (see picture below).

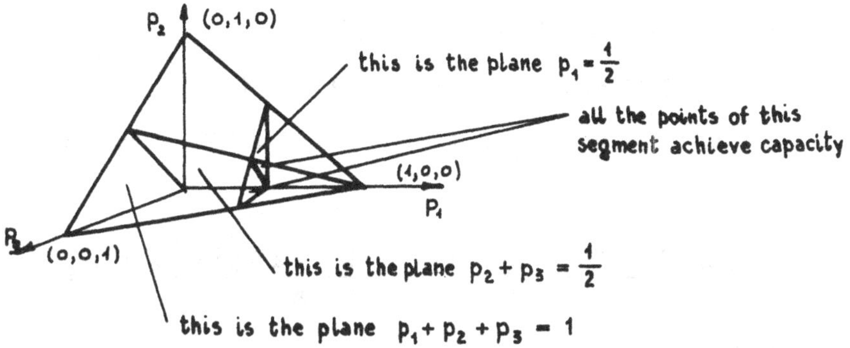

The fact that the stationary points of a convex function, in particular of $I(x;y)$, are really absolute maxima can be easily proved:

Theorem: suppose the function

$$\varphi(p_1, p_2, \ldots, p_n) \triangleq \varphi(\bar{p}) \quad ,$$

is convex on the set $S = \{\bar{p} : p \geq 0, \ \sum_1^n i\, p_i = 1\}$, so that

$$\varphi\left[(1-a)\bar{p}_1 + a\bar{p}_2\right] \geq (1-a)\,\varphi(\bar{p}_1) + a\,\varphi(\bar{p}_2) \qquad (2.12)$$

for each pair $\bar{p}_1, \bar{p}_2$ belonging to S. Now assume that $\varphi(\bar{p})$ is continuously differentiable for $\bar{p} > \bar{0}$ (i.e. for $p_i > 0$, $i = 1,..,n$) and that $\bar{p}^* = (p_1^*, \ldots, p_n^*)$ is a point in S such that

$$\left.\frac{\partial \varphi(\bar{p})}{\partial p_i}\right|_{\bar{p}=\bar{p}^*} = 0 \quad , \qquad (i = 1, 2, \ldots, n) \ . \qquad (2.13)$$

We claim that the function φ restricted to S has an absolute maximum in $\bar{p} = \bar{p}^*$.

In fact suppose there is a $\bar{p}' \in S$ such that:

$$\varphi(\bar{p}') > \varphi(\bar{p}^*) \quad , \tag{2.14}$$

and consider the points of the segment (entirely belonging to S, which is a convex set) joining $\bar{p}^*$ to $\bar{p}'$:

$$(1-a)\,\bar{p}^* + a\bar{p}', \qquad 0 \leq a \leq 1 \quad . \tag{2.15}$$

Using the convexity relation (2.12'), we get at once:

$$\frac{\varphi[(1-a)\bar{p}^* + a\,\bar{p}'] - \varphi(\bar{p}^*)}{a} \geq \frac{(1-a)\,\varphi(\bar{p}^*) + a\,\varphi(\bar{p}') - \varphi(\bar{p}^*)}{a} =$$

$$= \varphi(\bar{p}') - \varphi(\bar{p}^*) > 0 \, , \qquad 0 \leq a \leq 1 \, , \tag{2.16}$$

where the assumption (2.14) has been used. Now the validity of eq.s in (2.13) implies that the directional derivative of φ is zero at $\bar{p}^*$ for all possible directions, which means that, for a tending to 0, the difference quotient on the left in eq.(2.16) must tend to zero too, contradicting the last inequality in (2.16). Therefore the point $\bar{p}^* \in S$ such that eq.s (2.13) are satisfied is an absolute maximum for $\varphi(\bar{p})$.

2. 3. Capacity of a Discrete Memoryless Channel with Square Nonsingular Channel Matrix.

Now we turn to the problem of computing the channel capacity C defined in eq. (1.30) in the case of a discrete memoryless channel having a square ($n \times n$) nonsingular channel matrix M . Under these hypotheses we are able to calculate the capacity in a closed-form expression [11] , provided we assume also that

$$\delta = \sum_{1}^{n}{}_{j} \, s_{ji} \, e^{-\sum_{1}^{n}{}_{i} s_{ji} H(y/x = x_i)} > 0 \quad . \qquad (2.17)$$

for i = 1,..., n, being s_{ij} the element in the i-th row and in the j-th column of M^{-1} :

<u>Theorem</u>: If (2.17) holds, then we get for the capacity the expression:

$$C = \log_e \sum_{1}^{n}{}_{j} e^{-\sum_{1}^{n}{}_{i} s_{ji} H(y/x = x_i)} \,{}^{*}, \qquad (2.18)$$

and one of the distributions which achieve capacity is given by:

$$p_i = e^{-C} \delta_i \qquad i = 1, \ldots, n \qquad (2.19)$$

---

* Here we assume the base of the logarithms to be e for the sake of simplicity in computing the derivatives. It is not difficult to switch back to the base 2, i.e. to pass from nats to bits: it is sufficient to multiply by $\log_2 e$.

To prove our statement, consider the function

$$I(x\,;y) = -\sum_{1}^{n}{}_{i}\, q_i \log q_i + \sum_{1}^{n}{}_{ij}\, \pi_{ij} \log q_{i/j}, \tag{2.20}$$

which coincides with the channel rate (1.29) in the domain S^*, and that we have to maximize $I(x;y)$ in S. Now we could think of applying the method of Lagrange multipliers, but nothing is said in that method about constraints of the type

$$p_i \geq 0 \tag{2.21}$$

and only the equation $\sum_{1}^{n}{}_{i}\, p_i = 1$ must be considered as a constraint. So we try to maximize the function:

$$I(x\,;y) + \lambda \sum_{1}^{n}{}_{i}\, p_i = H(y) - H(y/x) + \lambda \sum_{1}^{n}{}_{i}\, p_i \tag{2.22}$$

and hope that eventually the solution meets conditions (2.21).

Keeping in mind eq.s (1.5) we compute the derivatives with respect to p_i of the function in (2.22):

$$\begin{aligned}
\frac{\partial H(y)}{\partial p_i} &= -\frac{\partial}{\partial p_i}\sum_{1}^{n}{}_{j}\, q_j \log q_i = -\frac{\partial}{\partial p_i}\sum_{1}^{n}{}_{j}\Big\{\sum_{1}^{n}{}_{k}\, p_k q_{j/k}\Big\}\log\Big\{\sum_{1}^{n}{}_{k}\, p_k q_{j/k}\Big\} = \\
&= -\Big\{\sum_{1}^{n}{}_{j}\, q_{j/i}\log q_j + \sum_{1}^{n}{}_{j}\Big[\Big(\sum_{1}^{n}{}_{k}\, p_k q_{j/k}\Big)\frac{q_{j/i}}{\sum_{1}^{n}{}_{k}\, p_k q_{j/k}}\Big]\Big\} = \\
&= -1 - \sum_{1}^{n}{}_{j}\, q_{j/i}\log q_j \,; \qquad (2.23)\\
\frac{\partial H(y/x)}{\partial p_i} &= \frac{\partial}{\partial p_i}\Big\{\sum_{1}^{n}{}_{j}\, p_j H(y/x = x_j)\Big\} = H(y/x = x_i)\,;\\
&\frac{\partial}{\partial p_i}\lambda\sum_{1}^{n}{}_{j}\, p_j = \lambda\,;
\end{aligned}$$

---

* Remark that the function in (2.20) is a function of the p_i only, since $q_j = \sum_{1}^{n}{}_{i}\, p_i q_{j/i}$ and the $q_{j/i}$ are given.

so that the equations:

$$\frac{\partial}{\partial p_i}\left[I(x;y)+\lambda\sum_1^n{}_j\, p_j\right]=0\,, \qquad (i=1,2,\ldots,n)\,, \tag{2.24}$$

become

$$-1-\sum_1^n{}_j\, q_{j/i}\log q_j - H(y/x=x_i)+\lambda=0\,, \quad (i=1,2,\ldots,n)\,. \tag{2.25}$$

Now, since $\sum_1^n{}_j\, q_{j/i} = 1$, eq.s (2.25) may also be written in the following form:

$$\sum_1^n{}_j\left\{(\lambda-1)-\log q_j\right\}q_{j/i} = H(y/x=x_i)$$

or, in matrix form:

$$M\begin{bmatrix}\lambda-1-\log q_1\\ \lambda-1-\log q_2\\ \vdots\\ \lambda-1-\log q_n\end{bmatrix}=\begin{bmatrix}H(y/x=x_1)\\ H(y/x=x_2)\\ \vdots\\ H(y/x=x_n)\end{bmatrix} \tag{2.26}$$

Eq.s (2.26) together with condition $\sum_1^n{}_j\, p_j = 1$ constitute our system. Since M is assumed to be nonsingular, from (2.26) we get

$$\begin{bmatrix}\lambda-1-\log q_1\\ \lambda-1-\log q_2\\ \vdots\\ \lambda-1-\log q_n\end{bmatrix}=M^{-1}\begin{bmatrix}H(y/X=x_1)\\ H(y/X=x_2)\\ \vdots\\ H(y/X=x_n)\end{bmatrix}\,. \tag{2.27}$$

or also

$$1-\lambda+\log q_j=-\sum_1^n{}_i\, s_{ji}\, H(y/x=x_i)\,, \qquad (j=1,2,\ldots,n)\,, \tag{2.28}$$

whence:

$$q_j e^{1-\lambda}=e^{-\sum_1^n{}_i s_{ji} H(y/x=x_i)}\,, \qquad (j=1,2,\ldots,n)\,, \tag{2.29}$$

and since $\sum_1^n{}_i\, p_i = 1$ implies $\sum_1^n{}_j\, q_j = 1$, summing on both sides of eq. (2.29) with respect to j, we get:

$$e^{1-\lambda}=\sum_1^n{}_j\, e^{-\sum_1^n{}_i s_{ji} H(y/x=x_i)}\,, \tag{2.30}$$

whence:

$$1-\lambda=\log\sum_1^n{}_j\, e^{-\sum_1^n{}_i s_{ji} H(y/x=x_i)}\,. \tag{2.30'}$$

Now from eq. (1.5) we have $q_j = \sum_{1}^{n} {}_i \pi_{ij} =$ $= \sum_{1}^{n} {}_i p_i q_{j/i}$, i.e. in matrix form:

$$[q_1\ q_2 \ldots q_n] = [p_1\ p_2 \ldots p_n]\ M \ , \qquad (2.31)$$

whence:

$$[q_1\ q_2 \ldots q_n]\ M^{-1} = [p_1\ p_2 \ldots p_n] \ , \qquad (2.32)$$

i.e.:

$$p_i = \sum_{1}^{n} {}_j\, s_{ji}\, q_j \ , \qquad (2.33)$$

and substituting into (2.33) the expression for q_j which is derived from eq.s (2.29), we have:

$$p_i = e^{\lambda - 1} \sum_{1}^{n} {}_j\, s_{ji}\, e^{-\sum_{1}^{n} {}_i s_{ji} H(y/x = x_i)} \qquad (2.34)$$

which, according to our assumption expressed by eq. (2.17), is a strictly positive quantity for i = 1,....,n.

Thus eq.s (2.30') and (2.34) yield numbers $p_1, p_2, \ldots p_n$ and λ such that $p_i > 0$ and $\sum_{1}^{n} {}_i p_i = 1$ and such that they satisfy eq.s (2.24). But since $I(x;y) + \lambda \sum_{1}^{n} {}_i p_i$ is the sum of a convex function and of a linear function, it is itself convex (in a domain S of nonnegative numbers whose sum is unity) and in force of what we proved about maxima of convex functions, we may arrive at the conclusion that we have found an absolute maximum of the function $I(x;y) + \lambda \sum_{1}^{n} {}_i p_i$ over the domain S.

However our task was to find a maximum for the rate $I(x;y)$; now we claim that the solution we have found as expressed by eq.s (2.34) gives an absolute maximum for the channel rate $I(x;y)$. In fact, suppose that the actual capacity C

is greater than the value $\bar{I}(x;y)$ of the rate obtained in correspondence with the distribution given by (2.34), call it $\{\bar{p}_i\}$. Let $\{p_i^*\}$ be a distribution achieving capacity. Then of course we have:

$$I(x;y) + \lambda < C + \lambda = C + \lambda \sum_1^n{}_i\, p_i^* \quad , \tag{2.35}$$

but on the other hand, since $\{\bar{p}_i\}$ achieves an absolute maximum for $I(x;y) + \lambda \sum_1^n{}_i\, p_i$ over S , we have also:

$$C + \lambda \sum_1^n{}_i\, p_i^* \le \bar{I}(x;y) + \lambda \sum_1^n{}_i\, \bar{p}_i = \bar{I}(x;y) + \lambda \quad , \tag{2.36}$$

which contradicts (2.35).

Now to find the channel capacity explicitly we start from eq. (2.25) and multiply it by p_i ; then summing over i we obtain:

$$H(y) - H(y/x) = 1 - \lambda \quad , \tag{2.37}$$

i.e. $1 - \lambda$ is the channel capacity. Using eq. (2.30') it turns out that:

$$C = \log_a \sum_1^n{}_j\, e^{-\sum_1^n{}_i\, s_{ji} H(y/x = x_i)} \tag{2.38}$$

Now some remarks to conclude this part: suppose M is square and nonsingular, but that assuption (2.17) does not hold for all i's. Then the Lagrange method will provide zero or negative values for some of the numbers p_i in (2.34). This means of course that the maximum is on the boundary of S . Thus an "input reduction" becomes necessary, i.e. we must set some of the p_i's equal to zero in the expression of $I(x;y)$ and try to maximize it as a function of the remaining variables. This turns out to be a reduction in the number of inputs, but it does not imply in general a reduction in the number of outputs, and therefore the reduced channel matrix is no longer square, so that the previous procedure is not applicable.

2. 4. A General Theorem.

We want now to state a very concise and elegant theorem [22] concerning the capacity of a discrete memoryless channel. Although this theorem is not constructive, nevertheless it gives some insight in the way capacity is achieved by an input probability distribution.

The theorem is as follows:

The input probability distribution $\{p_1, p_2, \dots p_n\}$ achieves capacity for the channel characterized by the channel matrix (2.1) if and only if

$$I(x_i\,;y) = C \qquad \text{for all } i \text{ with } p_i > 0 \tag{2.39'}$$

$$I(x_i\,;y) \leq C \qquad \text{for all } i \text{ with } p_i = 0 \tag{2.39''}$$

being $I(x_i\,;y)$ the mutual information for i-th input averaged over the outputs:

$$I(x_i\,;y) = \sum_j q_{j/i} \log \frac{q_{j/i}}{\sum_k p_k\, q_{j/k}} \quad . \tag{2.40}$$

Furthermore, the constant C in eq.s (2.39) is the capacity of the channel.

For the proof we need a lemma on convex functions which generalizes the results of section 2-2. We state this result without proof and in a form suitable for our purposes (for the proof see e.g. [22], ch.4):

Lemma. Let $\varphi(\bar{p})$ be a convex function in the set S defined in section 2-2. Assume that the partial derivatives $\partial\varphi(\bar{p})/\partial p_i$ exist and are continuous on S possibly with $\lim_{p_i\to 0}\partial\varphi(\bar{p})/\partial p_i = +\infty$ for some index i . Then the conditions:

$$\frac{\partial\varphi(\bar{p})}{\partial p_i} = \lambda \qquad \text{all } i \text{ with} \qquad p_i > 0 \qquad (2.41')$$

$$\frac{\partial\varphi(\bar{p})}{\partial p_i} \leq \lambda \qquad \text{all } i \text{ with} \qquad p_i = 0 \qquad (2.41'')$$

for some value of λ are necessary and sufficient conditions on $\bar{p}$ for the maximization of φ over S .

According to the definition of capacity, we want to maximize the quantity

$$I(x\,;y) = \sum_{i,j} p_i\, q_{j/i} \log \frac{q_{j/i}}{\sum_k p_k\, q_{j/k}} \qquad , \qquad (2.42)$$

which is a convex function in the set S , as we have already proved. (We assume the logarithms are to the base e). As we we have already seen (cf.(2.23)) the partial derivatives of $I(x\,;y)$ are

$$\frac{\partial I(x;y)}{\partial p_i} = -\sum_j q_{j/i} \log \sum_k p_k\, q_{j/k} - H(y/x = x_i) - 1 = \qquad (2.43)$$

$$= I(x_i\,;y) - 1 \qquad .$$

which satisfy the conditions of the Lemma. It follows that the necessary and sufficient conditions on the probability vector $\bar{p}$ to maximize $I(x;y)$ are

$$\frac{\partial I(x;y)}{\partial p_i} = \lambda \qquad \text{for} \qquad p_i > 0 \quad , \qquad (2.44')$$

$$\frac{\partial I(x;y)}{\partial p_i} \leq \lambda \qquad \text{for} \qquad p_i = 0 \quad . \qquad (2.44'')$$

From eq.s (2.44), using (2.43) the thesis (2.39) follows provided we put $C = 1 + \lambda$. To prove that C is actually the capacity, multiply both sides of (2.39') by p_i and sum over those i's for which $p_i > 0$. On the left we get the maximum of $I(x;y)$ and on the right the constant C , which completes the proof.

Although this result is not constructive, as we have observed, it is very interesting. Eq.s (2.39) amount to say that the capacity is attained when each input has the same mutual information as any other. If this were not the case, then we could improve the performance of our channel by using more often those inputs with larger mutual information; this however would change the mutual information relative to each input and after a sufficiently large number of changes, all inputs will provide the same amount of information, except possibly a few inputs whose information content is so scarce that their probabilities are made vanish. Of course, in these conditions the mean value of the mutual information relative to one input coincides with the mutual information of each of the inputs being used with non zero probability, and therefore, with the channel capacity.

Part Three

Measures for the Amount of Information.

3. 1. Introduction.

We have seen that a convenient measure for the uncertainty associated to (or the information contained in) the finite scheme

$$A = \begin{pmatrix} a_1 & a_2 & \dots & a_n \\ p_1 & p_2 & \dots & p_n \end{pmatrix} \qquad \left(p_i \geq 0 , \quad \sum_1^n {}_i \, p_i = 1 \right) \tag{3.1}$$

is given by the Shannon entropy $H(A) = -\sum_1^n {}_i \, p_i \log p_i$. Since H(A) actually depends only on the finite discrete probability distribution, i.e. on the numbers $p_1 , p_2 , \dots ; p_n$ we shall often write

$$\mathcal{G} = (p_1 , p_2 , \dots , p_n) \text{ and } H(\mathcal{G}) \tag{3.2}$$

instead of (3.1) and H(A) respectively.

A great deal of work [12] , [13] , [14] has been devoted to give axiomatic foundations for the entropy:

$$H(\mathcal{G}) = - \sum_1^n {}_i \, p_i \log p_i = H(p_1 , \dots , p_n) \quad , \tag{3.3}$$

and various sets of postulates have been given.

We do not intend to give here a list of these postulate sets, or to give proofs about the sufficiency of these postulates. We limit ourselves to reproduce here the postulates suggested by Faddejew [13] .

i) H (p, 1-p) is continuous for $0 \leq p \leq 1$ and positive at least in one point;

ii) $H(P_1, P_2, \dots, P_n)$ is a symmetric function of the variables $P_1, \dots, P_n$;

iii) for $n \geq 2$ the following equality holds:

$$H(P_1, P_2, \dots, P_{n-1}, q_1, q_2) = H(P_1, P_2, \dots, P_{n-1}, P_n) + \qquad (3.4)$$
$$+ P_n H\left(\frac{q_1}{P_n}, \frac{q_2}{P_n}\right),$$

where $q_1 + q_2 = P_n$.

A kind of "normalization" requirement may also be added:

iv) $H(\frac{1}{2}, \frac{1}{2}) = 1$,

which simply defines the units in which entropy is measured.

It can be proved that postulates i) to iv) uniquely define the function $H(P_1, P_2, \dots, P_n)$ as having the form $H(\mathcal{P}) = -\sum_1^n {}_i P_i \log P_i$.

We remark that if $\mathcal{P} = (P_1, P_2, \dots, P_n)$ and $\mathcal{Q} = (q_1, q_2, \dots q_m)$ are two independent probability distributions and $\mathcal{P} * \mathcal{Q}$ is their direct product, i.e. the distribution consisting in the m.n numbers $P_i q_j$ ($i = 1, \dots, n$; $j = 1, \dots, m$), then eq. (3.4) implies that

$$H(\mathcal{P} * \mathcal{Q}) = H(\mathcal{P}) + H(\mathcal{Q}) . \qquad (3.5)$$

Eq. (3.5) expresses a kind of additivity for the entropy (compare with eq. (1.15)), which holds also for dependent schemes $\mathcal{P}$ and $\mathcal{Q}$ (cf. eq. (1.11)). Now observe that eq. (3.5) cannot replace postulate iii), because it is much weaker than eq. (3.4). In fact there are many quantities different from (3.3) which satisfy postulates i), ii), iv) and eq. (3.5). So do e.g. [15] , [16] the quantities:

$$H_\alpha(\mathcal{G}) = H_\alpha(p_1, p_2, \ldots, p_n) = \frac{1}{1-\alpha} \log\left(\sum_1^n{}_j \, p_j^\alpha\right), \quad (\alpha > 0;\ \alpha \neq 1) \qquad (3.6)$$

which were defined by Rényi and are called "entropies of (positive) order α" of the distribution $\mathcal{G}$.

Remark that

$$\lim_{\alpha \to 1} H_\alpha(\mathcal{G}) = H(\mathcal{G}) \quad . \qquad (3.7)$$

as it is easily seen, so that Shannon's entropy (3.3) is the limiting case for $\alpha \to 1$ of the entropy H . Therefore, (3.3) will be indicated by $H_1(\mathcal{G})$ and can be called the measure of order 1 of the entropy (uncertainty) of $\mathcal{G}$.

3. 2. Incomplete Schemes and Generalized Probability Distributions.

There are many practical cases where the result of an experiment is not always (i.e. with probability 1) observable, but it is observable only with a probability ≤ 1 . This remark has led to the introduction of the "generalized probability distributions".

Consider a probability space $[\Omega, \mathcal{A}, P]$ and let $\xi = \xi(\omega)$ be a r.v. defined for $\omega \in \Omega_1$, where $\Omega_1 \in \mathcal{A}$ and $P(\Omega_1) \leq 1$.

If $P(\Omega_1) = 1$ we call ξ an "ordinary" (or "complete") r.v.; if instead $0 < P(\Omega_1) < 1$, we call ξ an "incomplete" r.v.

In any case ξ is a generalized r.v., and its probability distribution is called a generalized probability distribution.

In the simplest case ξ corresponds to a finite scheme as in (3.1), where now the quantity $P(\Omega_1)$ coincides with

$$W(\mathcal{G}) = \sum_1^n i\, p_i \quad , \qquad (3.8)$$

called the "weight" of the distribution; $W(\mathcal{G})$ satisfies the condition:

$$0 < W(\mathcal{G}) \leq 1 \qquad (3.9)$$

If the equality holds on the right in eq. (3.9) the distribution $\mathcal{G}$ is complete; otherwise it is incomplete.

A set of postulates has been given [15] for the entropy H($\mathcal{G}$) in case $\mathcal{G}$ is any set of $n \geq 1$ nonnegative numbers (P_1 , P_2 , ... , P_n) such that $0 < \sum_1^n i\, p_i \leq 1$. Let us call Δ the ensemble of all such sets; then we have:

Postulate 1: H ($\mathcal{G}$) is a symmetric function of the numbers P_1 , P_2 , ... , P_n ;

Postulate 2: If $\{p\} \in \Delta$ consists of the single probability p , then H $(\{p\})$ is a continuous function of p in the interval $0 < p \leq 1$ (the continuity is not required for p = 0);

Postulate 3: $H\left(\left\{\frac{1}{2}\right\}\right) = 1$;

Postulate 4: If $\mathcal{G} \in \Delta$, $\mathcal{Q} \in \Delta$ we have $H(\mathcal{G} * \mathcal{Q}) = H(\mathcal{G}) + H(\mathcal{Q})$.

Now in order to state Postulate 5 let $\mathcal{G}$ = (P_1 , ... , P_n) and

$\mathcal{Q} = (q_1, \ldots, q_n)$ be two generalized distributions such that $W(\mathcal{G}) + W(\mathcal{Q}) \leq 1$, and put

$$\mathcal{G} \cup \mathcal{Q} = (p_1, \ldots, p_n, q_1, \ldots, q_n) \; ; \tag{3.10}$$

if $W(\mathcal{G}) + W(\mathcal{Q}) > 1$, $\mathcal{G} \cup \mathcal{Q}$ is not defined. Then we have:

Postulate 5: If $\mathcal{G} \in \Delta$ and $\mathcal{Q} \in \Delta$ and $W(\mathcal{G}) + W(\mathcal{Q}) \leq 1$, then

$$H(\mathcal{G} \cup \mathcal{Q}) = \frac{W(\mathcal{G})\, H(\mathcal{G}) + W(\mathcal{Q})\, H(\mathcal{Q})}{W(\mathcal{G}) + W(\mathcal{Q})} \; . \tag{3.11}$$

We are now in a position to state and prove the following

Theorem: If $H(\mathcal{G})$ is defined for $\mathcal{G} \in \Delta$ and satisfies Postulates 1 to 5, then $H(\mathcal{G}) = H_1(\mathcal{G})$, where

$$H_1(\mathcal{G}) = \frac{-\sum_1^n p_i \log p_i}{\sum_1^n p_i} \; .$$

Proof. Set $h(p) = H(\{p\})$ for $0 < p \leq 1$ (cf. Postulate 2). From Postulate 4 it follows:

$$h(pq) = h(p) + h(q) \qquad (0 < p \leq 1 \, ; \; 0 < q \leq 1) \tag{3.12}$$

and from Postulate 3 $h(\frac{1}{2}) = 1$. As it is well known, this is sufficient to conclude that

$$h(p) = -\log p \; ^{*} .$$

Now, if $\mathcal{G}_1, \mathcal{G}_2, \ldots, \mathcal{G}_n$ are incomplete distributions such that $\sum_1^n W(\mathcal{G}) \leq 1$, then applying Postulate 5 we easily get:

$$H(\mathcal{G}_1 \cup \mathcal{G}_2 \cup \ldots \cup \mathcal{G}_n) = \frac{W(\mathcal{G}_1)\, H(\mathcal{G}_1) + W(\mathcal{G}_2)\, H(\mathcal{G}_2) + \ldots + W(\mathcal{G}_n)\, H(\mathcal{G}_n)}{W(\mathcal{G}_1) + W(\mathcal{G}_2) + \ldots + W(\mathcal{G}_n)} \; , \tag{3.13}$$

---

* Of course here the logarithms are taken to the base 2.

and since any generalized distribution $\mathcal{G} = (p_1, p_2, \ldots, p_n)$ may be written in the form:

$$\mathcal{G} = \{p_1\} \cup \{p_2\} \cup \ldots \cup \{p_n\} \quad , \tag{3.14}$$

the completion of the proof follows from (3.13) and (3.14).

Now looking back at (3.3) we see that Shannon's entropy is an average, i.e. it is the mean value of the quantity $\log 1/p_i$, which is actually the entropy of the generalized distribution $\{p_i\}$, and coincides with the self-information of an event having probability p_i.

We remark at this point that in eq. (3.11) an arithmetic mean has been postulated. The question may be asked whether other kinds of mean may be used. In general the mean value of the numbers $x_1, x_2, \ldots, x_n$ with respect to the weights $w_1, w_2, \ldots, w_n$ ($w_i > 0$, $\sum_{1}^{n} {}_i w_i = 1$) has the form:

$$g^{-1}\left\{\sum_{1}^{n} {}_i w_i \, g(x_i)\right\} \quad , \tag{3.15}$$

being $g(x) = y$ any strictly monotonic and continuous function. The form of (3.15) corresponding to our case of generalized distributions is the following:

$$H(\mathcal{G} \cup \mathcal{Q}) = g^{-1}\left\{\frac{W(\mathcal{G})\, g[H(\mathcal{G})] + W(\mathcal{Q})\, g[H(\mathcal{Q})]}{W(\mathcal{G}) + W(\mathcal{Q})}\right\} \quad . \tag{3.16}$$

On the ground of these remarks one is led to replace Postulate 5 by the following

<u>Postulate 5'</u>: There exists a strictly monotonic and continuous function $g(x)$ such that if $\mathcal{G} \in \Delta$ and $\mathcal{Q} \in \Delta$. with $W(\mathcal{G}) + W(\mathcal{Q}) \leq 1$

then eq. (3.16) holds.

Since we feel that Postulate 4 is very important (we want indeed that the uncertainty, or information, expressed by any entropy possesses the property of additivity), the following question naturally arises: which are the functions $g(x)$ appearing in (3.16) which are compatible with Postulate 4?

It has been proved [16] that the only compatible choices for $g(x)$ in Postulate 5' are

$$g_1(x) = ax + b \quad , \tag{3.17}$$

in which case Postulate 5' reduces to Postulate 5 and the required function turns out to be Shannon's entropy:

$$H_1(\mathcal{G}) = \frac{-\sum_1^n{}_i\, p_i \log p_i}{\sum_1^n{}_i\, p_i} \quad ; \tag{3.18}$$

and

$$g_\alpha(x) = 2^{-(\alpha-1)x} \quad , \qquad (\alpha > 0\,,\ \alpha \neq 1) \tag{3.19}$$

in which case Postulates 1,2,3,4 and 5' characterize the entropy of order α :

$$H_\alpha(\mathcal{G}) = \frac{1}{1-\alpha} \log\left[\frac{\sum_1^n{}_i\, p_i^\alpha}{\sum_1^n{}_i\, p_i}\right] \tag{3.20}$$

Clearly, if $\mathcal{G}$ is a complete distribution (i.e. if $\sum_1^n{}_i\, p_i = 1$), then (3.18) reduces to (3.3) and (3.19) to (3.6).

It is also readily seen that

$$\lim_{\alpha\to 1} H_\alpha(\mathcal{G}) = H_1(\mathcal{G}) \quad . \tag{3.21}$$

3. 3. Characterization of Shannon's Entropy among the Entropies of Positive Order.

Now the problem arises as which is the property characterizing the entropy (3.3) (entropy of order 1, say) among the entropies of order α . In fact both H_1 and H_α satisfy postulates 1, 2, 3, 4 and 5' (and are the only functions which satisfy).

As we have already observed, the additivity property expressed either by Postulate 4 or by eq. (3.5) is, as far as Shannon's entropy is concerned, a consequence of eq. (3.4) in Faddejew's axiom iii); but it is weaker than eq. (3.4) itself. So that it is evident that while both H_α ($\mathcal{G}$) ($\alpha \neq 1$) and H_1 ($\mathcal{G}$) satisfy the additivity property, only H_1 ($\mathcal{G}$) possesses the stronger property (3.4), in force of the uniqueness theorem stated and proved by Faddejew.

In other words among all functions of the form

$$\mathcal{H}(\mathcal{G}) = g^{-1}\left(\sum_1^n{}_i \, p_i \, g(-\log p_i)\right) \quad , \tag{3.22}$$

being $g(x)$ a continuous strictly monotonic function such that the function

$$f(x) = \begin{cases} x\, g(-\log x) & \text{for } x \neq 0 \\ 0 & \text{for } x = 0 \end{cases} \tag{3.23}$$

is strictly convex, both H_1 ($\mathcal{G}$) and H_α($\mathcal{G}$) ($\alpha > 0, \alpha \neq 1$) satisfy the additivity condition

$$\mathcal{H}(\mathcal{G} * \mathcal{Q}) = \mathcal{H}(\mathcal{G}) + \mathcal{H}(\mathcal{Q}) \quad . \tag{3.24}$$

for independent schemes $\mathcal{P}$ and $\mathcal{Q}$; moreover $H_1(\mathcal{P})$ and $H_\alpha(\mathcal{P})$ are the only functions satisfying (3.24). It can be shown that the strict convexity of $f(x)$ in (3.23) together with the continuity and monotonicity of $g(x)$ in (3.22) is equivalent to a kind of monotonicity of $\mathcal{H}(\mathcal{P})$, i.e. to the very important inequality

$$\mathcal{H}(\mathcal{Q}/\mathcal{P}) \le \mathcal{H}(\mathcal{Q}) \quad , \tag{3.25}$$

which we have proved for $H_1(\mathcal{P})$ (cf.(1.28)), and which seems to be fully justified from an intuitive point of view.

But, although $H_\alpha(\mathcal{P})$ and $H_1(\mathcal{P})$ both satisfy eq. (3.24) (independent distributions) and eq. (3.25), only $H_1(\mathcal{P})$ satisfies the more stringent condition

$$\mathcal{H}(\mathcal{P} * \mathcal{Q}) = \mathcal{H}(\mathcal{P}) + \mathcal{H}(\mathcal{Q}/\mathcal{P}) \tag{3.26}$$

for non-independent distributions $\mathcal{P}$ and $\mathcal{Q}$, where

$$\mathcal{H}(\mathcal{Q}/\mathcal{P}) = g^{-1}\left\{\sum_1^n{}_i\, p_i \sum_1^m{}_j\, q_{j/i}\, g(-\log q_{j/i})\right\} \quad , \tag{3.27}$$

which particularizes into

$$H_1(\mathcal{Q}/\mathcal{P}) = -\sum_1^n{}_i\, p_i \sum_1^m{}_j\, q_{j/i} \log q_{j/i} \quad , \tag{3.28}$$

$$H(\mathcal{Q}/\mathcal{P}) = \frac{1}{1-\alpha} \log\left\{\sum_1^n{}_i\, p_i \sum_1^m{}_j\, q_{j/i}^\alpha\right\} , \qquad \alpha \neq 1 \tag{3.29}$$

respectively. So (3.28) does satisfy eq.(3.26) while (3.29) can not satisfy it for $\alpha \neq 1$, as follows from Faddejew's uniqueness theorem.

However the property expressed by eq. (3.26) holds also for $H_\alpha(\mathcal{B})$ provided we modify the form of $H_\alpha(\mathcal{Q}/\mathcal{B})$ in the following way:

$$H'_\alpha(\mathcal{Q}/\mathcal{B}) = \frac{1}{1-\alpha} \log\left\{\sum_1^n{}_i \frac{p_i^\alpha}{\sum_1^n{}_i p_i^\alpha} \sum_1^m{}_j q_{j/i}^\alpha\right\} .$$

To conclude consider the three conditions:

$$\mathcal{H}(\mathcal{B}) = \mathcal{H}(p_1, p_2, \ldots, p_n) \leq \mathcal{H}\left(\frac{1}{n}, \ldots, \frac{1}{n}\right); \quad (3.30)$$

$$\mathcal{H}(p_1, p_2, \ldots p_n) = \mathcal{H}(p_1, p_2, \ldots, p_n, 0); \quad (3.31)$$

$$\mathcal{H}(\mathcal{B} * \mathcal{Q}) = \mathcal{H}(\mathcal{B}) + \mathcal{H}(\mathcal{Q}/\mathcal{B}) . \quad (3.32)$$

Condition (3.31) is satisfied by all the functions of the form (3.22) with the assumptions relative to (3.23); condition (3.30) is satisfied by all previous functions having a $g(x)$ which is strictly convex; condition (3.32) is satisfied only by H_1 and by H_α in the case of independent distributions and only by H_1 in the general case [16] .

This is a very nice characterization of Shannon's entropy.

3. 4. Pragmatic Approach to Uncertainty Measures.

So far we have dealt with an axiomatic characterization for a suitable measure of the uncertainty (or information) connected with a (finite) r.v. As we have already said, much work has been done in this direction, and still being done. We must however observe that a different approach can also be thought of, i.e. a pragmatic approach.

In the axiomatic approach, starting from a certain number of reasonable properties the information should possess, one formulates a certain number of axioms and finds out all the functions which meet these conditions, possibly restricting the class of such functions by imposing extra requirements. In the pragmatic approach instead, particular problems are coped with and the quantities coming out in the solution of these problems are claimed to be suitable measures of uncertainty (or information) on the basis that they simply worked.

It is clear that these two points of view are strongly interrelated, because on one hand a full justification for the functions found by an axiomatic approach lies in their capability of solving real problems of information theory (which thus provide a test for any axiomatic approach); on the other hand, among these functions which come out from the solution of particular problems only those which can serve to solve a rather broad class of problems can provide suitable measures for uncertainty or information, and in their turn can be arrived at in an axiomatic way.

We only mention two particular problems which are of special importance and are brilliantly coped with in terms of information theory methods and tools: the problem of coding the information generated by a source, and the problem of "random search". The latter problem is approximately stated as follows [17] , [18] : suppose we want to find out a particular object, say x , among the n objects of a set S_n . We are allowed to perform m independent experiments, each consisting in dividing at random S_n into k subsets $A_1, \ldots, A_n$. We know the probability that any element of S_n belongs to the class A_i (and this without taking into account what happens to the other elements). The result of each experiment is that we are informed which one among the classes A_i contains the unknown element x. The problem then consists in finding out the element x with a probability exceeding $1 - \varepsilon$ (ε an arbitrary number between 0 and 1), and this by means of as few subdivisions of S_n as possible.

Now it has been shown that Shannon's entropy and Rényi's entropies of positive order play a fundamental role in this kind of problems. Thus this is a good example of a pragmatic approach to the entropy functions.

3.5. Informational Entropy and Informational Energy.

As we have already remarked, any kind of uncertainty measure is a mean value (through the function $g(x)$, cf. eq.(3.22)) of the quantity log 1/p, which can of course be considered as a r.v. having the distribution $\mathcal{G} = (p_1, p_2, \ldots, p_n)$. Every such mean value is a number, and therefore provides only a partial information about the distribution $\mathcal{G}$. Consequently, since there are many ways of averaging, there are many possible forms for the uncertainty measure, as we have already seen, each of which gives some information about $\mathcal{G}$. On the other hand, one might think of averaging in (3.22) functions other than log 1/p; e.g. one could simply take p, which is itself a r.v. with the distribution $\mathcal{G}$. In this way, choosing for $g(x)$ the identity function, we get the "information energy" of $\mathcal{G}$, which has been thoroughly studied by Onicescu and others [19], [20], [21], [22]:

$$E(\mathcal{G}) = \sum_1^n{}_i\, p_i^2 \quad . \tag{3.33}$$

$E(\mathcal{G})$ has many interesting properties, although it does not satisfy an additive condition of the kind of (3.24), but rather a multiplicative property.

Of course, by comparing (3.33) with (3.20) we realize that

$$H_2(\mathcal{G}) = -\log E(\mathcal{G}) = \log 1/E(\mathcal{G}) \quad . \tag{3.34}$$

If we indicate by $M(\eta) = \sum_1^n{}_i\, \eta_i p_i$ the mean value of the r.v. η taking on the values η_i with probability p_i (i = 1, 2, ..., n) and by $L(x)$ the function log 1/x, we clearly have:

$$H_1(\mathcal{G}) = M(L(\pi)) \tag{3.35}$$

$$H_2(\mathcal{G}) = L\,(M(\pi)) = L\,(E(\mathcal{G})) \quad , \tag{3.36}$$

being π the r.v. assuming the values p_i with probability p_i.

3. 6. Uncertainty Functions for Countable Schemes.

Now we attempt to define the uncertainty functions also for non-finite schemes. The first step consists in passing from finite schemes:

$$\begin{pmatrix} x_1 & x_2 & \dots & x_n \\ p_1 & p_2 & \dots & p_n \end{pmatrix} \qquad \left(p_i \geq 0\,;\ \sum_1^n {}_i\, p_i = 1\right), \tag{3.37}$$

to countable schemes:

$$\begin{pmatrix} x_1 & x_2 & \dots & x_n & \dots \\ p_1 & p_2 & \dots & p_n & \dots \end{pmatrix} \qquad \left(p_i \geq 0\,;\ \sum_1^\infty {}_i\, p_i = 1\right). \tag{3.38}$$

Then the entropy of a r.v. ξ associated to the scheme (3.38) is defined as:

$$H(\xi) = -\sum_1^\infty {}_i\, p_i \log p_i \ , \tag{3.39}$$

provided the series on the right converges*. The information energy of ξ is defined as:

$$E_0(\xi) = \sum_1^\infty {}_i\, p_i^2 \ . \tag{3.40}$$

Remark that the series in (3.40) always converges, since $p_i^2 \leq p_i \leq 1$ and $\sum_1^\infty {}_i\, p_i^2 \leq \sum_1^\infty {}_i\, p_i = 1$, so that any discrete scheme possesses an informational energy.

At this point, to put things into a more general form, it is advisable to refer to a probability space

---

* The series may also be divergent, in which case we say that ξ has no entropy.

$(\Omega\ ,\mathcal{A}\ ,P\ )$, and to attach an entropy or an energy to every partition of Ω into measurable sets.

Consider first finite partitions:

$$\Omega=\bigcup_{i=1}^{n}A_i \qquad (\ A_i\in\mathcal{A}\ ,\ i=1,\dots,n\ ;\ A_iA_j=\emptyset \text{ if } i\neq j\ )\ .$$

Now, letting $P(A_i)=p_i$, we say that the entropy and the energy of the partition $\{A_i\}$ are given as usual by $-\sum_1^n{}_i\, p_i \log p_i$ and $\sum_1^n{}_i\, p_i^2$.

It is very easily seen that such a partition $\{A_i\}$ is equivalent to a "simple r.v." (a r.v. is said to be simple when it takes on only a finite number of different values a_i), say $\xi\,(\omega)$, defined in the space Ω as follows:

$$\xi(\omega)=\sum_1^n{}_i\, a_i\, \varphi_{A_i}(\omega)\ , \tag{3.42}$$

being $\varphi_{A_i}(\omega)$ the characteristic function of the measurable set A_i :

$$\varphi_{A_i}(\omega)=\begin{cases}1 & \text{if } \omega\in A_i\\ 0 & \text{if } \omega\notin A_i\end{cases}\ , \tag{3.43}$$

and being $a_i\neq a_j$ for $i\neq j$. In other words:

$$A_i=\{\omega:\xi(\omega)=a_i\}\ . \tag{3.44}$$

Then the entropy or the energy of the partition $\{A_i\}$ may be considered also as the entropy or the energy of the simple r.v. $\xi(\omega)$ in (3.42). It is apparent that the values of these entropy and energy functions do not depend on the values a_i , as long as they are different from each other, but only on the numbers p_i .

The extension of these considerations to the countable partitions and corresponding r.v.s is now straightforward.

3. 7. The Continuous Case. The Intensity of the Informational Energy.

Now consider a (real) r.v. $\xi(\omega)$ whatsoever, and let $\mathcal{B}_1$ be the Borel field on the real line R_1 . Consider in R_1 the half-open intervals α of the form $[a,b)$; of course, since $\alpha = [a, b) \in \mathcal{B}_1$ the sets

$$A_\alpha = \{\omega : \xi(\omega) \in \alpha\} \tag{3.45}$$

belong to $\mathcal{A}$, and form a σ -algebra $\mathcal{A}_f$ in $\mathcal{A}$. If the r.v. $\xi(\omega)$ is discrete and corresponds to the scheme (3.37) or (3.38), then we have:

$$A_{(-\infty,a)} = \{\omega : \xi(\omega) \in (-\infty, a)\} = \bigcup_{j:a_j \leq a} \{\omega : \xi(\omega) \in (-\infty, a_i)\} =$$
$$= \bigcup_{j:a_j \leq a} A_{(-\infty,a_i)} \in \mathcal{A}, \quad \forall a \in R_1 , \tag{3.46}$$

so that $\mathcal{A}_f$ is generated by a countable number of sets belonging to $\mathcal{A}$. The distribution function of such a discrete r.v.:

$$Prob\{A_{(-\infty,a)}\} = \Xi(a) \tag{3.47}$$

is a piecewise continuous function, actually a "step function" (possibly with a countable number of steps), possessing the usu al properties of such functions.

In the general case [23] the distribution function $\Xi(a)$ of a r.v. $\xi(\omega)$ is given by the superposition of a step function and of a continuous function, so that the σ-algebra $\mathcal{A}_f$ is made up by a certain number (at most countable) of atoms* having probabilities $p_1, p_2, \ldots, p_n, \ldots$ and by a set Ω_0, having positive probability, p_0, which is infinitely divisible; i.e. given any two nonnegative numbers p_0' and p_0'', with $p_0'+p_0''=p_0$, then Ω_0, can be partitioned into two parts Ω_0', Ω_0'' having probabilities p_0' and p_0'' respectively. Of course:

$$\sum_{1}^{\infty} {}_i\, p_i + p_0 = 1 \qquad (0 < p_0 < 1)\,. \tag{3.48}$$

Now let us forget about the atoms, and consider only the infinitely divisible set Ω_0; this amounts to omit the jumps in the distribution function $\Xi(a)$ of $\xi(\omega)$, and therefore $\Xi(a)$ becomes a continuous function with the properties:

$$\lim_{a\to-\infty} \Xi(a) = 0 \tag{3.49}$$

$$\lim_{a\to+\infty} \Xi(a) = p_0 > 0\,.$$

The corresponding r.v. will be continuous, and we want to define the energy and the entropy of such a r.v. First of all we suppose $\Xi(a)$ possesses a derivative at every real

---

* An atom is a measurable set $A \in \mathcal{A}$ such that $p(A) > 0$ and for any $B \in \mathcal{A}$ and $B \subset A$ we have either $p(B) = 0$ or $p(B) = p(A)$.

point a^{*}, which we call its "density" and indicate by $\rho(a)$, so that:

$$\Xi(b) - \Xi(a) = \int_a^b \rho(x)\, dx \quad . \tag{3.50}$$

---

* This assumption amounts to neglect the possible "continuous singular" part of $\Xi(a)$ and to consider only its "absolutely continuous" part. Let us recall [24] that a countably additive set function $\Phi(E)$ defined over a measure space $(\mathcal{A}, \mu)$ ($\mathcal{A}$ a σ-ring of subsets of a set Ω , μ a measure) is said to be:

i) "continuous" if it is defined and equal to zero on every set E containing a single point and such that $\mu(E) = 0$;

ii) "absolutely continuous" if it is defined and equal to zero on every set E having measure zero;

iii) "singular" if it is concentrated on a set E_0 of measure zero, i.e. if $\Phi(E) = 0$ for every measurable set $E \subset \Omega - E_0$;

iv) "discrete" if it is concentrated on a set E of measure zero containing no more than countably many points.

Any countably additive set function $\Phi(E)$ defined over the measure space $(\mathcal{A}, \mu)$ may be expressed in the form:

$$\Phi(E) = A(E) + S(E) + D(E) \quad , \tag{+}$$

where $A(E)$ is absolutely continuous, $S(E)$ is continuous and singular, $D(E)$ is discrete. Moreover:

$$A(E) = \int_A \varphi(x)\, \mu(dx) \quad , \tag{++}$$

Now we consider a sequence of step functions $\{\Xi_n(a)\}$, converging from below to $\Xi(a)$:

$$\Xi_n(a) \uparrow \Xi(a), \text{ as } n \to \infty . \tag{3.51}$$

Each of these step functions $\Xi_n(a)$ is the distribution function of a simple r.v. $\xi_n(\omega)$ defined in Ω_0 (for the sake of simplicity assume from now on that $\Omega_0 \equiv \Omega$), and so determines a (finite) partition in Ω , since Ω is infinitely divisible: $\Pi^{(n)} = \{A_j^{(n)}\}_{j=1}^{m_n}$.

It is not difficult to see that we may choose a monotonic sequence of partitions $\Pi^{(n)} = \{A_j^{(n)}\}_{j=1}^{m_n}$ (n = 1, 2, ...), in the sense that $\Pi^{(n+1)}$ is a refinement of $\Pi^{(n)}$ (i.e. each set in $\Pi^{(n+1)}$ is contained in one set of $\Pi^{(n)}$). Now the energy $E(\Pi^{(n)})$ is defined in the usual way: $\sum_{j=1}^{m_n} [\text{Prob } (A_j^{(n)})]^2$, and it is immediately seen that if $\Pi^{(n+1)}$ refines $\Pi^{(n)}$, then:

$$E(\Pi^{(n+1)}) \leq E(\Pi^{(n)}) \quad . \tag{3.52}$$

Given any sequence $(\varepsilon_n)_{n\in N}$ of real numbers such that $\varepsilon_n < 1$ (n=1, 2 ....) and

---

* where the function $\varphi(x)$ is μ -summable on every $E \in \mathcal{A}$. The representation (+) is unique and so is $\varphi(x)$ in (++) up to a set of measure zero.

Now for every set $A \in \mathcal{B}_1$ of the form [a,b) consider the set function $\Phi(A) = \Xi(b^+) - \Xi(a^-)$ (i.e. $\Xi(x)$ is the "generating function" or "distribution function" of Φ); then what is said about Φ may be restated in an obvious form about Ξ .

$$\varepsilon_n \downarrow 0 \ , \tag{3.53}$$

we may choose in the sequence $\{\Pi^{(n)}\}$ a subsequence $\{\Pi^{(tn)}\}$ such that

$$\text{prob}\ (\ A_j^{(tn)}\ ) < \varepsilon_n\ (n=1,2,\ldots\ldots). \tag{3.54}$$

Now:

$$E(\Pi^{(tn)}) = \sum_1^{m_{tn}}{}_j \left[\text{Prob}\,(A_j^{(tn)})\right]^2 < \varepsilon_n \sum_1^{m_{tn}}{}_j \text{Prob}\,(A_j^{(tn)}) = \varepsilon_n \ ,$$

since $\sum_1^{m_{tn}}{}_j \text{Prob}\ (\ A_j^{(tn)}\ ) = 1$, and then (3.53) implies

$$E(\Pi^{(n)}) \downarrow 0 \ , \tag{3.55}$$

so that $E(\Xi) = \lim_{n\to\infty} E(\Xi_n)$ is not a good definition for the energy of the continuous r.v. $\xi(\omega)$ corresponding to $\Xi(a)$.

As regards entropy, by the same procedure we find that it tends to ∞. In fact from (3.53) you get $\frac{1}{\varepsilon_n} \uparrow \infty$ and since $\log x$ is a monotonic function, from (3.54) we have:

$$\log \frac{1}{\text{Prob}\,(A^{(tn)})} > \log \frac{1}{\varepsilon_n} \ ,$$

so that

$$H(\Pi^{(tn)}) = \sum_1^{m_{tn}}{}_j \text{Prob}\,(A_j^{(tn)}) \log \frac{1}{\text{Prob}\,(A_j^{(tn)})} > \log \frac{1}{\varepsilon_n} \ ,$$

and therefore $H(\Pi^{(n)}) \uparrow \infty$.

Once more $H(\Xi) = \lim_{n\to\infty} H(\Xi_n)$ is not a good definition for the entropy of the continuous r.v. $\xi(\omega)$ corresponding to $\Xi(a)$.

These facts make us change our definition of entropy and energy in the continuous case, and make us introduce a sort of entropy or energy "intensity", in the following way. Consider again the sequence of simple r.v.s $\xi_n(\omega)$ and the corresponding distributions $\Xi_n(a)$. Put:

$$\text{Prob}\,(A_{j+1}^{(n)}) = \Xi_n(a_{j+1}^{(n)}) - \Xi(a_j^{(n)}) = \Delta\,\Xi_n(a_j^{(n)}) \quad , \tag{3.56}$$

and therefore for the energy corresponding to the partition

$\Pi^{(n)} = \{A_j^{(n)}\}_{j=1}^{m_n}$ we have:

$$E^{(n)} = E(\Pi^{(n)}) = \sum_0^{m_n-1}{}_j \left\{\text{Prob}\,(A_{j+1}^{(n)})\right\}^2 = \sum_1^{m_n}{}_j \frac{\Delta\,\Xi_n(a_j^{(n)})}{\Delta\, a_j^{(n)}}\, \Delta\,\Xi(a_j^{(n)})\, \Delta\, a_j^{(n)} \quad , \tag{3.57}$$

where $\Delta\, a_j^{(n)} = a_{j+1}^{(n)} - a_j^{(n)}$, $a_j^{(n)}(j = 1, 2, \ldots, m_n)$ being the points of the real axis where the jumps of $\Xi_n(a)$ are located.

In eq. (3.57) we may choose the $\Delta\, a_j^{(n)}$ of constant value, say $\Delta\, a_j^{(n)} = \Delta\, a^{(n)}$ $(j = 1, \ldots, m_n)$, whence we get:

$$\frac{E^{(n)}}{\Delta a^{(n)}} = \sum_1^{m_n}{}_j \frac{\Delta\,\Xi^{(n)}(a_j^{(n)})}{\Delta\, a^{(n)}}\, \Delta\,\Xi^{(n)}(a_j^{(n)}) \quad . \tag{3.58}$$

Now we let n tend to the infinity, making so $\Delta\, a^{(n)}$ tend to zero, together with $E^{(n)}$, as we have already seen:

$$\lim_{n\to\infty} \Delta\, a^{(n)} = \lim_{n\to\infty} E^{(n)} = 0 \tag{3.59}$$

It may happen that the limit

$$\lim_{n\to\infty} \frac{E^{(n)}}{\Delta\, a^{(n)}} \tag{3.60}$$

exists, in which case, as it is seen from the right-hand side of eq. (3.58), it reads as follows:

$$\int_{R_1} \rho(x)\, \Xi(dx) = \int_{R_1} \rho^2(x)\, dx \quad ,$$

where we have taken into account the definition of the density $\rho(x)$ of the distribution $\Xi(x)$ (see eq. (3.50)).

If the quantity (3.61) exists, then owing to (3.60), it is called the energy "intensity" of the r.v. $\xi(\omega)$ or of its distribution $\Xi(x)$, appearing on the right in eq.(3.51).

Of course the following properties hold:

$$\rho(x) \geq 0 , \qquad \int_{R_1} \rho(x)\, dx = 1 , \qquad (3.62)$$

unless an atomic (discrete) part exists for $\xi(\omega)$, in which case the second equation in (3.62) is modified as follows: $\int_{R_1} \rho(x)\, dx = p_0 < 1$, according to equation (3.18).

As an example, consider a Gaussian r.v., having the density:

$$\rho(x) = \frac{1}{\sigma\sqrt{2\pi}} \exp\left(-\frac{x^2}{2\sigma^2}\right) , \qquad (3.63)$$

which satisfies the two conditions in (3.62). In this case, the intensity of the information energy, according to eq. (3.61), is given by:

$$\frac{1}{\sigma^2 2\pi} \int_{R_1} e^{-\frac{x^2}{\sigma^2}} dx = \frac{1}{2\pi\sigma^2}\, \sigma\sqrt{\pi} = \frac{1}{2\sigma\sqrt{\pi}} \quad \left(\text{Remember that } \int_{R_1} e^{-x^2} dx = \sqrt{\pi}\right) . \qquad (3.64)$$

It is thus apparent that the energy intensity is inversely proportional to σ, which is the variance, and therefore increases when the concentration of the Gaussian curve increases. In the limit for $\sigma \to \infty$ (uniform distribution = greatest uncertainty), the energy intensity goes to zero.*

---

* This property is shared by the energy of a finite scheme,

Suppose $A \subset R_1$ is the real set where $\rho(x) \geq a > 0$; then putting

$$I_a \doteq \int_A \rho^2(x)\,dx \; ; \quad I = \int_{R_1} \rho^2(x)\,dx \; ; \quad \int_A \rho(x)\,dx = \Delta_A \Xi(x) \, ,$$

we get:

$$I_a = \int_A \rho^2(x)\,dx \geq a \int_A \rho(x)\,dx = a\,\Delta_a \Xi(x) \quad ,$$

so that:

$$\Delta_A \Xi(x) = \mathrm{Prob}\,\{\omega : \xi(\omega) \in A\} \leq \frac{I_a}{a} \leq \frac{I}{a} \quad ,$$

which is an upper bound for the probability of the set $\{\omega : \xi(\omega) \in A\}$.

So we have seen that, given a r.v. $\xi(\omega)$, its informational energy is made up of two parts: i) the part relative to the atoms, having the form $\sum_1 p_i^2$, which is always defined (indeed ≤ 1), and vanishes only when there is no atom or when there are countably many equally likely atoms; and ii) the part relative to the absolutely continuous part (we do not consider here the continuous singular part for the distribution function of $\xi(\omega)$), having the form $\int_{R_1} \rho^2(x)\,dx$, which can also diverge.

---

* which -contrary to the entropy- achieves its minimum value 1/n when the n cases are equally likely, i.e. when the uncertainty is greatest.

3. 8. The Continuous Case, Entropy.

As regards the entropy, we proceed in the same way, again omitting the atomic part. Instead of considering the sequence of the mean values of the logarithms, which goes to the infinity, we consider the mean value of the quantity

$$-\log \frac{\Delta \Xi (a_j^{(n)})}{\Delta a_j^{(n)}} , \tag{3.65}$$

(cf. (3.57)), i.e. the quantity:

$$\bar{H}_n = -\sum_{1}^{m_n}{}_j \, \Delta \Xi (a_j^{(n)}) \log \frac{\Delta \Xi (a_j^{(n)})}{\Delta a_j^{(n)}} . \tag{3.66}$$

When n goes to the infinity, the right-hand term of eq. (3.66) may tend to a limit, which has the form:

$$-\int_{R_1} \log \rho (x) \, \Xi (dx) = -\int_{R_1} \rho (x) \log \rho (x) \, dx . \tag{3.67}$$

Then, when the integral in (3.67) exists, i.e. is finite, we assume it as the entropy of the corresponding (absolutely) continuous r.v. $\xi (\omega)$.
Remark however that when $\rho (x) = \delta (x)$, then eq. (3.67) yields ∞ instead of zero (the δ distribution corresponds to zero uncertainty). This is the price to be paid for this "renormalization" of entropy in the continuous case.

As an example, consider again the Gaussian density function $\rho (x) = \frac{1}{\sigma \sqrt{2\pi}} \exp \left(- \frac{x^2}{2\sigma^2}\right)$; we get for its entropy intensity:

$$-\int_{R_1} \frac{1}{\sigma\sqrt{2\pi}}\, e^{-\frac{x^2}{2\sigma^2}} \log\left\{\frac{1}{\sigma\sqrt{2\pi}} - \frac{x^2}{2\sigma^2}\right\} dx = 1 + \log \sigma\sqrt{2\pi} .$$

This entropy intensity clearly increases as σ increases, i.e. when the concentration of the Gaussian curve decreases, and goes to the infinity when σ does, i.e. when the distribution tends to the uniform distribution.

3. 9. An Interpretation for the Intensity of the Informational Energy.

Now we give an interpretation of the expression (3.61), which defines the intensity of the informational energy [21] . We shall see that if an absolutely continuous distribution function $f(x)$ is given in the interval $I = \{|x| \leq K\}$,then the quantity

$$\int_{-K}^{K} \rho^2(x)\, dx , \tag{3.68}$$

where $\rho(x) = \frac{d f(x)}{dx}$ is a measure of the amount by which $f(x)$ differs in I from the uniform distribution $f_u(x) = \frac{1}{2K} x + \frac{1}{2}$.

We consider a r.v. ξ , such that $|\xi| \leq K$ having a distribution function $f(x)$, absolutely continuous, and therefore possessing a density $\rho(x)$, such that:

$$f(x) = \int_{-\infty}^{x} \rho(t)\, dt = \int_{-K}^{x} \rho(t)\, dt . \tag{3.69}$$

Now, consider a sequence of "simple" r.v.s $\{\xi^{(n)}\}$, defined in the following way [25] :

$$\xi^{(n)} = \frac{[n\xi]}{n}^{*} . \qquad (3.70)$$

or equivalently:

$$\xi^{(n)} = \frac{k}{n} \qquad \text{for} \qquad \frac{k}{n} \leq \xi < \frac{k+1}{n}$$

$$(k = -[nK], -[nK]+1, \ldots, -1, 0, 1, \ldots, [nK]-1) .$$

The distribution function of $\xi^{(n)}$ in (3.70) is given by:

$$\begin{aligned} f^{(n)}(x) &= \int_{-K}^{x} \sum_{k=-\infty}^{+\infty} \left[\left\{ f\left(\frac{k+1}{n}\right) - f\left(\frac{k}{n}\right) \right\} \delta\left(t - \frac{k+1}{n}\right) \right] dt = \\ &= \sum_{k=-[nK]}^{[nx]-1} \left\{ f\left(\frac{k+1}{n}\right) - f\left(\frac{k}{n}\right) \right\} = f\left(\frac{[nx]}{n}\right) - f\left(-\frac{[nK]}{n}\right) , \end{aligned} \qquad (3.71)$$

and its informational energy is given by:

$$E^{(n)} = \sum_{k=-[nK]}^{[nK]-1} \left\{ f\left(\frac{k+1}{n}\right) - f\left(\frac{k}{n}\right) \right\}^{2} . \qquad (3.72)$$

It is immediately seen that $\lim_{n \to \infty} f^{(n)}(x) = f(x)$. Moreover as usual:

$$\lim_{n \to \infty} E^{(n)} = 0 . \qquad (3.73)$$

In fact, from (3.72) we have:

---

* $[\cdot]$ denotes the integral part.

$$0 \leq E^{(n)} \leq \sup_k \left\{ f\left(\frac{k+1}{n}\right) - f\left(\frac{k}{n}\right) \right\} \cdot \sum_{k=-[nK]}^{[nK]-1} \left\{ f\left(\frac{k+1}{n}\right) - f\left(\frac{k}{n}\right) \right\} =$$

$$= \sup_k \left\{ f\left(\frac{k+1}{n}\right) - f\left(\frac{k}{n}\right) \right\} \cdot \left\{ f\left(\frac{[nK]}{n}\right) - f\left(\frac{-[nK]}{n}\right) \right\} , \tag{3.74}$$

and the first factor in the last term of (3.74) tends to zero as $n \to \infty$, because of the continuity of $f(x)$, while the second factor remains bounded as $n \to \infty$ (more precisely it tends to* 1). So we find once more the result expressed by (3.55). Now remark that for any n fixed we have $2[nK]$ possible values for $\xi^{(n)}$ and from the basic properties of the informational energy [20], we have:

$$E^{(n)} \geq \frac{1}{2[nK]} , \tag{3.75}$$

i.e. also:

$$\lim_{n \to \infty} E^{(n)} 2[nK] \geq 1 , \tag{3.76}$$

which tells us that $E^{(n)}$ goes to zero (for $n \to \infty$) not faster than $\frac{1}{2[nK]}$.

Now put nK in the place of $[nK]$, and suppose that $\rho(x)$ is continuous in $I = \{|x| \leq K\}$. The limit in (3.76) becomes then:

$$\lim_{n \to \infty} E^{(n)} \cdot 2nK = 2K \lim_{n \to \infty} n \sum_{k=-nK}^{nK-1} \left\{ f\left(\frac{k+1}{n}\right) - f\left(\frac{k}{n}\right) \right\}^2 \leq$$

$$\leq 2K \lim_{n \to \infty} n M_n \cdot \sum_{k=-nK}^{nK-1} \left\{ f\left(\frac{k+1}{n}\right) - f\left(\frac{k}{n}\right) \right\} = 2K \lim_{n \to \infty} n \cdot M_n , \tag{3.77}$$

---

* Actually $\lim_{n \to \infty} \pm \frac{[nK]}{n} = \pm K$, so that in the limit $\pm [nK] \cong \cong \pm nK$.

being

$$M_n = \sup_k \left\{ f\left(\frac{k+1}{n}\right) - f\left(\frac{k}{n}\right) \right\} \quad . \tag{3.78}$$

In force of the continuity of $f(x)$, for any k and n fixed, we have:

$$f\left(\frac{k+1}{n}\right) - f\left(\frac{k}{n}\right) = \frac{1}{n} f'(x_k^{(n)}) = \frac{1}{n} \rho(x_k^{(n)}) \quad , \tag{3.79}$$

where $\frac{k}{n} \leq x_k^{(n)} \leq \frac{k+1}{n}$ and therefore:

$$M_n = \sup_k \frac{1}{n} \rho(x_k^{(n)}) = \frac{1}{n} \sup_k \rho(x_k^{(n)}) \ .$$

Now, putting $\sup_k \rho(x_k^{(n)}) = D_n$ and $\lim_{n\to\infty} D_n = D$, we get

$$\lim_{n\to\infty} E^{(M)} 2nK \leq 2KD \quad ; \tag{3.80}$$

remark that $D = \lim_{n\to\infty} D_n$ is the maximum of $\rho(x)$ in the (closed) interval $[-K,K]$, where $\rho(x)$ is continuous.

From (3.76) and (3.80) we may conclude that if the limit $\lim_{n\to\infty} E^{(n)} 2nK$ exists, then $E^{(n)}$ goes to zero as 1/n. But that limit exists, since, taking into account (3.79):

$$\lim_{n\to\infty} 2nK\, E^{(n)} = \lim_{n\to\infty} 2nK \sum_{-nK}^{nK-1}{}_k \left\{ \frac{1}{n} \rho(x_k^{(n)}) \right\}^2 =$$

$$= 2K \lim_{n\to\infty} \sum_{-nK}^{nK-1}{}_k \frac{1}{n} \rho^2(x_k^{(n)}) = 2K \int_{-K}^{K} \rho^2(x)\, dx \ ,$$

or also

$$\lim_{n\to\infty} nE^{(n)} = \int_{-K}^{K} \rho^2(x)\, dx \quad . \tag{3.81}$$

We have seen a careful derivation of the intensity of the informational energy in the case the distribution $f(x)$ of the r.v. ξ has a continuous derivative. Now for the expression $I_f = 2K \int_{-K}^{K} \rho^2(x)\, dx$ we have from (3.76) and (3.80) the following inequalities:

$$1 \leq I_f \leq 2kD^{*} \qquad (3.82)$$

The lower bound in (8.32) corresponds to the equality in (3.75), which occurs when $\xi^{(n)}$ has $2[nK]$ equally likely outcomes, and then** corresponds to the uniform distribution $f_u(x) = \frac{1}{2K} x + \frac{1}{2}$ (i.e. constant density $\rho_u(x) = 1/2\,K$) in the interval $-K, K$.

The greater I_f is with respect to 1, the more "different" $f(x)$ is from the linear distribution $f_u(x)$.

3. 10. Dimension of a Probability Distribution and d - dimensional Entropy.

The use of the sequence of the r.v.s $\{\xi^{(n)}\}$ appearing in (3.70) is very useful also with respect to further considerations about the entropy of a continuous distribution function. We have seen that given a r.v. ξ with absolutely continuous distribution function $f(x)$, the expression

$$H(\xi) = -\int_{-\infty}^{+\infty} \rho(x) \log \rho(x)\, dx \quad . \qquad (3.67)$$

---

* Of course this imposes also the bound on D: $D \geq \frac{1}{2K}$.

** Since it must be valid for any n .

may be taken as its entropy, being $\rho(x)$ the density of $f(x)$, provided the integral exists. Now let us make a distinction between the entropy of a discrete (finite or countable) r.v. and the entropy of a continuous r.v. and the entropy of a continuous r.v.; precisely we put:

$$-\sum_{1}^{\infty} {}_i\, p_i \log p_i = H_0(\xi) = \text{entropy for a discrete r.v.;} \quad (3.83)$$

$$-\int_{-\infty}^{\infty} \rho(x) \log \rho(x)\, dx = \quad \text{entropy for a continuous r.v.;} \quad (3.83')$$

provided the series or the integral converge. From the definition, it follows that some of the properties of $H_1(\xi)$ are similar to those of $H_0(\xi)$, but other properties are completely different. Consider for instance the uniform density defined as follows:

$$\rho(x) = \begin{cases} \frac{1}{\alpha} & \text{for } 0 \leq x \leq \alpha \\ 0 & \text{otherwise.} \end{cases}$$

Then the entropy of the corresponding distribution $f(x) = \int_{-\infty}^{x} \rho(z)\, dz$ is given by $-\int_{0}^{\alpha} \frac{1}{\alpha} \log \frac{1}{\alpha}\, dx = \log \alpha$ and if $\alpha \leq 1$, then $H_1(f) \leq 0$. It is perhaps apparent since now that the entropies of discrete distributions and those of continuous distributions are not comparable, i.e. are quantities of different kind. To support this statement, consider the real numbers in $[0,1]$, and write them in binary form: $\sum_{1}^{\infty} {}_i \frac{a_i}{2^i}$ where the coefficients a_i are 0 or 1. To specify any number η chosen at random in $[0,1]$ we could give its coefficients. Since η is chosen at random it seems sensible to assume that the values zero and one are assumed by the coefficients with the same probability $\frac{1}{2}$, and that the coefficients are independent of each other; thus each coefficient provides one bit of information, and the information provided by η is the sum of the information provided by its coefficients, i.e. $H_0(\eta) = \infty$. But on the other hand, by the previous argument, $H_1(\eta) = \log 1 = 0$.

It may be shown [25] [26] that besides H_0(zero-dimensional entropy) and H_1 (one-dimensional entropy) infinitely many entropies $H_d (0 \leq d \leq 1)$ may be defined (d-dimensional entropies), for different kinds of r.v.s. The entropies H_d may assume any value from $-\infty$ to $+\infty$ when $0 < d \leq 1$, while when $d = 0$ only positive values are taken on.

We may introduce an ordering among the entropies of the various r.v.s, or equivalently of the various probability distributions, in the following way: if ξ_1 and ξ_2 are two r.v.s having dimension d_1 and d_2 respectively, then if $d_1 < d_2$ we say that the entropy of ξ_1 is less than the entropy of ξ_2 independently of their actual values. If $d_1 = d_2$, then we compare the values of the entropies in the usual way.

Now suppose a bounded r.v. ξ whatsoever is given, and consider again the sequence $\{\xi^{(n)}\}$ of r.v.s defined by eq. (3.63):

$$\xi^{(n)} = \frac{[n\xi]}{n} \quad . \tag{3.63}$$

If $|\xi| < K$, $\xi^{(n)}$ can assume at most $2nK$ different values, so that $H_0(\xi^{(n)}) \leq \log 2nK$.

Now if the limit

$$\lim_{n\to\infty} \frac{H_0(\xi^{(n)})}{\log n} = d(\xi) \quad . \qquad (3.84)$$

exists, then of course $d(\xi) \le 1$, and the number $d(\xi)$ defined by (3.84) is called the "dimension" of (the distribution function of) the r.v. ξ (remark that when $n \to \infty$, $\xi^{(n)} \uparrow \xi$). If the limit does not exist, then the dimension of ξ is not defined.

If the dimension $d(\xi)$ does exist, we define the quantity:

$$H_{d(\xi)} = \lim_{n\to\infty} \{ H_0(\xi^{(n)}) - d(\xi) \log n \} \quad , \qquad (3.85)$$

if the limit exists, as the $d(\xi)$ -dimensional entropy of ξ . If the limit does not exist, then this entropy is not defined.

The definitions (3.84) and (3.85) are consistent with what we have said about discrete and continuous r.v.s. In other words: if ξ is a discrete r.v. taking on the values (which must be distinct) x_i with probabilities p_i , then $d(\xi) = 0$ and $H_0(\xi) = -\sum_{1}^{\infty} {}_i\, p_i \log p_i$, provided the series converges; if ξ is an absolutely continuous r.v. with density $\rho(x)$, and if $H_0([\xi])$* is finite, then $d(\xi) = 1$ and $H_1(\xi) = -\int_{-\infty}^{\infty} \rho(x) \log \rho(x)\, dx$, provided the integral exists.

---

* $H([\xi])$ is the zero-dimensional entropy of the discrete r.v. $[\xi]$, where as usual $[\cdot]$ denotes the integral part.

We now consider a r.v. ξ such that $H_0([\xi])$ is finite and the distribution function $f(x)$ of ξ has the form

$$f(x) = (1-\alpha)\, f_0(x) + \alpha\, f_1(x) \qquad (0 < \alpha < 1)\,, \qquad (3.86)$$

where $f_0(x)$ is a discrete distribution $\{p_i\}$ and $f_1(x)$ is an absolutely continuous distribution with density $\rho(x)$. Then the dimension $d(\xi)$ of ξ is given by

$$d(\xi) = \alpha \quad .$$

If moreover both the series $-\sum_{i=1}^{\infty} p_i \log p_i$ and the integral $-\int_{-\infty}^{\infty} \rho(x) \log \rho(x)\, dx$ are convergent, then:

$$H_\alpha(\xi) = (1-\alpha) \sum_{i=1}^{\infty} p_i \log \frac{1}{p_i} + \alpha \int_{-\infty}^{\infty} \rho(x) \log (1/\rho(x))\, dx +$$
$$+ \alpha \log 1/\alpha + (1-\alpha) \log \frac{1}{1-\alpha} \quad .$$

As it is well known from measure theory [24] [27], any set function generated by a probability distribution function is the sum of three set functions, one discrete, one continuous and singular, one absolutely continuous. In the case of probability distribution functions we have:

$$f(x) = a\, f_1(x) + b f_2(x) + c\, f_3(x) \;,$$

where a, b and c are nonnegative numbers whose sum is unity. So far we have been treating the case where $b = 0$ (i.e. the continuous singular part is absent), and it seems [26] that the cases where $b > 0$ are particularly complicated to handle.

To close, we just mention that the results which we have exposed till now can be extended to r-dimensional random vectors (r = 2, 3,.....), i.e. to probability distributions in the Euclidean spaces E_r, and a definition of dimension can be given which coincides with the ordinary (geometrical) definition in the case of absolutely continuous distribution functions.

Part Four

The Source - Coding Theorem.

4. 1. Sources and Stationary Sources.

We shall consider an information source Q as a random process (r.p.), and we shall restrict ourselves to the discrete r.p.s. Such a r.p. is made up by the doubly infinite sequences ξ of the form:

$$\xi = \ldots x_{-n}\ x_{-n+1} \ldots x_{-1}\ x_0 x_1 \ldots x_{n-1}\ x_n \ldots \qquad (4.1)$$

where the symbols ("letters") x come from a finite source alphabet A having a (>1) elements. Let S be the totality of the sequences of the form (4.1).

Particular interest among the subsets of S deserve those subsets built as follows: fix k positions (instants) in ξ : t_n , t_{n+1} , ..., t_{n+k-1} and k letters (not necessari ly distinct) α_n, α_{n+1} ,...., α_{n+k-1} ; now the event "the source Q generates the letter α_i at time t_i ($n \leq i \leq n+ + k-1$)" is made up by all those sequences ξ in which $x_i = \alpha_i$. Any such set will be called "(thin) cylinder of order k" and will be indicated by ${}_n\omega^{(k)}$.

Let Ω be the smallest field containing all the cylinders of finite order, and let P be a probability on Ω . It turns out that the source is completely defined by the set

$$\{S, \Omega, P\}$$.

It is apparent that for any fixed values n and k the cylinders ${}_n\omega^{(k)}$ form a partition of S into a^k disjoint sets.

Now consider the following translation transformation T for every $\xi \in S$:

$$\xi = (\ldots x_{-n} \ldots x_{-1}\ x_0 x_1 \ldots x_n \ldots) \xrightarrow{T} T\xi = (\ldots y_{-n} \ldots y_{-1} y_0 y_1 \ldots \ldots y_n \ldots), \tag{4.2}$$

where:

$$y_j = x_{j-1} \quad \text{any } j .$$

Naturally T induces a mapping Φ between the cylinders: to a cylinder ${}_n\omega_i^{(k)}$ corresponds a cylinder $\Phi(\,{}_n\omega_i^{(k)})$ (which in general differs from ${}_n\omega_i^{(k)}$). Consequently in S the partition $\{{}_n\omega_i^{(k)}\}_{i=1}^{a^k}$ is transformed into the partition $\{\Phi(\,{}_n\omega_i^{(k)})\}_{i=1}^{a^k}$.

If we assume that P is invariant with respect to T , i.e. if

$$P(\,{}_n\omega_i^{(k)}) = P(\Phi(\,{}_n\omega_i^{(k)})) \qquad (i = 1, \ldots, a^k) \tag{4.3}$$

for any finite k , then we say that the source $Q = \{S, \Omega, P\}$ is "stationary".

If a source is stationary, then we may forget about n in the definition of cylinders. More precisely, in S , TS , T^2S ,....the cylinders corresponding to the same n-tuple (α_1 , $\alpha_2 \ldots \alpha_n$) of letters are different, but have all the same probability.

4. 2. Separation of the Source - Encoder from the Channel - Encoder.

With reference to the block "coder" in the figure of section 1.2, we wish to point out that it is directly connected on one side with the Source, and on the other with the Channel. (Analogously the block "decoder" is connected with the Channel and with the Destination). This entails that their design should take into account the characteristics of both the Channel and the Source. One could think of splitting the coder (and the decoder) into a "Source Encoder" and a "Channel Encoder", as it is shown in the figure referred to above.

Apparently such a separation has the advantage that the two parts of the coder can be designed independently of each other, provided only a matching between the two is possible, i.e. provided in the points marked by a star in the following picture:

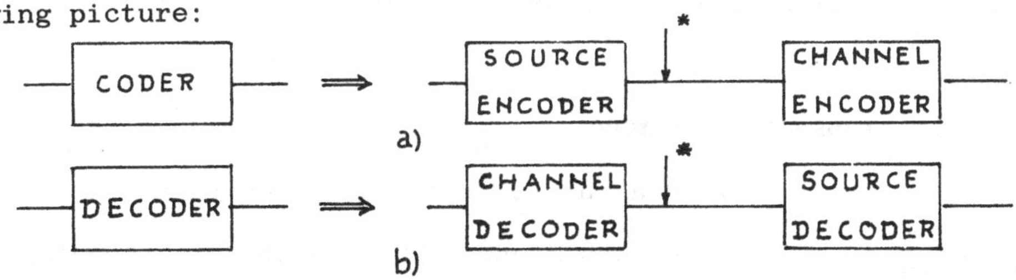

an interface-like form of the information is present. This possibility is by no means obvious [3] , [22] , but indeed this separation can be accomplished, and without reducing - under very broad conditions - the performances of the communication system.

This constitutes one of the most important results in Information Theory, and allows us to treat the source-coding and the channel-coding problems separately.

Moreover this separation makes it possible that different sources feed the same channel and that different channels are fed by the same source.

One practical and very often used way of performing this separation is to make the source encoder represent the source output (star in part a) of the above illustration) by a sequence of binary digits. It is this sequence of binary digits which is then presented to the channel encoder, that knows nothing about the way it was generated.

For simplicity's sake in the following formal description we shall assume that the source output is directly transmissible through the channel (i.e. we shall neglect the encoding problems), so that the source output alphabet coincides with the channel input alphabet.

4. 3. The Transmission Channel.

Now we assume that the stationary source $Q_A = \{S_A, \Omega_A, P_A\}$ * feeds a channel $\mathcal{C}$, which is defined as follows. Two finite alphabets are available, one is the source

---

* Here the subscript A refers to the source alphabet A .

alphabet A , which has a letters, and the other is an "output" alphabet B having b letters. (Of course both A and B are greater than 1). Starting from the family S_B of the sequences:

$$\eta = \ldots y_{-n}\; y_{-n+1} \ldots y_{-1}\; y_0\; y_1 \ldots y_n \ldots \qquad (4.14)$$

of elements drawn from B we may build a field Ω_B , in the same way as Ω_A was built. Furthermore over Ω_B a family of probability measures $\{{}_\xi\Pi_B\}_{\xi \in S_A}$ is given in such a way that:

$$Q_B(\xi) = \{ S_B , \Omega_B , {}_\xi\Pi_B \} \quad . \qquad (4.15)$$

is a source (cf. 4-1) for any sequence ξ of S_A . Then we say that the channel is made up by the triplet:

$$\{ A , {}_\xi\Pi_B , B \} \qquad (4.16)$$

where the existence of the fields Ω_A and Ω_B is implicit.
A channel defined as in (4.16) is said to be "stationary" if for the transformation T (cf.(4.2)) we have

$$_{T\xi}\Pi_B\,(T_B\,\omega) = {}_\xi\Pi_B\,({}_B\omega) \quad , \qquad (4.17)$$

being ${}_B\omega$ an element of Ω_B whatsoever, and $T_B\omega$ the cylinder obtained from ${}_B\omega$ after applicating T .

Now let

$$\xi^{(i)} = \ldots . \alpha_{-1}^{(i)} \ \alpha_0^{(i)} \ \alpha_1^{(i)} \ldots$$
$$\xi^{(j)} = \ldots . \alpha_{-1}^{(j)} \ \alpha_0^{(j)} \ \alpha_1^{(j)} \ldots \qquad (4.18)$$

be two particular sequences in S_A and let

$$\eta^{(i)} = \ldots . \beta_{-1}^{(i)} \ \beta_0^{(i)} \ \beta_1^{(i)} \ldots$$
$$\eta^{(j)} = \ldots . \beta_{-1}^{(j)} \ \beta_0^{(j)} \ \beta_1^{(j)} \ldots \qquad (4.19)$$

be the two sequences of S_B into which the channel transforms $\xi^{(i)}$ and $\xi^{(j)}$. It may happen that if in (4.18) $\alpha_k^{(i)} = \alpha_k^{(j)}$ ($k \leq n$; n an integer), then

$$_{\xi^{(i)}}\Pi_B(_B\Psi_i) = {}_{\xi^{(j)}}\Pi_B(_B\Psi_j) \ , \qquad (4.20)$$

where

$$_B\Psi_i = \{\eta : y_k = \beta_k^{(i)}\} \qquad (k \leq n)$$
$$_B\Psi_j = \{\eta : y_k = \beta_k^{(j)}\} \qquad (k \leq n) . \qquad (4.21)$$

If condition (4.20) is met for any n and for any choice of the sequences in (4.18), then the channel is "non anticipating".

Now let $\mathcal{C}$ be a nonanticipating channel, and, with reference to (4.18) and (4.19), assume that if $\alpha_k^{(i)} = \alpha_k^{(j)}$ for $n - m \leq k \leq n$ (m , n integers, $m \geq 0$), then (4.20) holds, where now $_B\Psi_i$, $_B\Psi_j$ are respectively the cylinders in

Ω_B :

$$ {}_B\Psi_i = {}_B\omega_i = \left\{ \eta \; : \; y_k = \beta_k^{(i)} \right\} ; $$

$$ {}_B\Psi_j = {}_B\omega_j = \left\{ \eta \; : \; y_k = \beta_k^{(j)} \right\} \qquad (n - m \le k \le n) . $$

The least (integer) value of m for which the above-decribed property is valid, is called the "memory" of the channel $\mathcal{C}$.

Finally let $\xi = \ldots\ldots \alpha_{-1} \quad \alpha_0 \quad \alpha_1 \ldots\ldots$ be a sequence generated by the source Q_A and $\eta = \ldots\ldots \beta_{-1} \quad \beta_0 \quad \beta_1 \ldots$ be the corresponding sequence generated by the source $Q_B(\xi)$. Consider two blocks* in: $\alpha_p \quad \alpha_{p+1} \ldots\ldots \alpha_q \; (q \ge p)$ and $\alpha_r \; \alpha_{r+1} \ldots\ldots \alpha_s \; (s \ge r)$ and assume that $u = r - q$ is nonnegative.

If there exists some value of u such that:

$$ {}_\xi\Pi_B \left({}_B\omega_p^{(q-p)} \cap {}_B\omega_r^{(s-r)} \right) = {}_\xi\Pi_B \left({}_B\omega_p^{(q-r)} \right) {}_\xi\Pi_B \left({}_B\omega_r^{(s-r)} \right) , ^{**} \qquad (4.22) $$

let $\bar{u}$ be the minimum such value. Then the source $Q_B(\xi)$ is said to be "$\bar{u}$ - independent". Remark that $\bar{u}$ depends on the particu-

---

* Given any discrete (finite or infinite) sequence ξ of letters, a "block" is any finite nonempty subsequence of consecutive letters of ξ .

** The subscripts "p" and "q" in eq. (4.22) indicate the first position in the two blocks.

lar sequence ξ , and that if m is the memory of the channel, then $\bar{u}(\xi) \geq m$ for every $\xi \in S_A$.

4. 4. The Flows Associated with the Channel.

Supposing the transmission through the channel $\mathcal{C}$ is instantaneous, we consider the letter x_t input to $\mathcal{C}$ at time t and the letter y_t output from $\mathcal{C}$ at the same time, i.e. the letters generated by Q_A and by $Q_B(\xi)$ respectively at time t . Consider the pair

$$z_t = (x_t , y_t) \tag{4.23}$$

and the sequences ζ of the form:

$$\zeta = \ldots z_{-n} \; z_{-n+1} \ldots z_{-1} \; z_0 \; z_1 \ldots z_{n-1} z_n \ldots , \tag{4.24}$$

which consist of letters taken from the finite alphabet $D = A \times B$ having $a \cdot b$ letters. Of course from (4.23) and (4.24) we see that it makes sense to put $\zeta = (\xi , \eta)$.

Let S_D be the totality of the sequences (4.24), in which as usual we may consider the cylinders of finite order k :

$$D^{\omega^{(k)}} = \left\{ \zeta : \delta_{t_n} \ldots \delta_{t_{n+k-1}} = \pi_D \right\} , \tag{4.25}$$

where π_D is a point in $D^k = D \times D \times \ldots \times D$ and the $\delta = (\alpha, \beta)$ are fixed letters in D (couples of letters one from A , one from B). The cylinders (4.25) generate a field Ω_D , which is the Cartesian product, in the sense of Probability Theory [29] , of the fields Ω_A and Ω_B *.

Now we introduce in Ω_D an additive function P_D which is defined on the cylinders in the following way:

$$P_D({}_D\omega) = \int_{{}_A\omega} {}_\xi\Pi_B({}_B\omega)\; P_A(d\,\xi)\;, \tag{4.26}$$

being $({}_A\omega\,,\,{}_B\omega) = {}_D\omega$. It is not difficult to verify [20] , [30] that P_D is a probability measure.

Now, as we have said in 4-1, the cylinders of order k constitute in S_A , S_B and S_D partitions of the order a^k, b^k, $d^k = a^k\, b^k$ respectively. Now it is easily seen that if ${}_A\omega_i \in \Omega_A$ and ${}_B\omega_j \in \Omega_B$ are cylinders, then, letting ${}_D\omega_{ij} = ({}_A\omega_i\,,\,{}_B\omega_j)$ we have also:

$${}_D\omega_{ij} = {}_A\omega_i \times {}_B\omega_j \qquad \left(i = 1, \ldots, a^k;\; j = , \ldots, b^k\right) \tag{4.27}$$

---

* Precisely each cylinder ${}_D\omega \in \Omega_D$ is a pair $({}_A\omega\,,\,{}_B\omega)$ of cylinders, with ${}_A\omega \in \Omega_A$, ${}_B\omega \in \omega_B$. Moreover among the elements of Ω_D those of the form

$$({}_A\omega\,,\,S_B)$$

generate a subfield $\Phi_A \subset \Omega_D$ which is isomorphic to Ω_A , and those of the form

$$(S_A\,,\,{}_B\omega)$$

generate a subfield $\Phi_B \subset \Omega_D$ isomorphic to Ω_B .

Consider in Ω_D the following unions of cylinders:

$$ {}_D\Theta_j = \bigcup_{i=1}^{a^k} {}_D\omega_{ij} \, , \qquad {}_D\varphi_i = \bigcup_{j=1}^{b^k} {}_D\omega_{ij} \qquad (4.28)$$

$$(j = 1, \ldots, b^k) \qquad\qquad (i = 1, \ldots, a^k)$$

which are of the forms given in the footnote at pag. 97.

For the probability of the sets in (4.28) we have of course with reference to (4.27):

$$P_D({}_D\vartheta_j) = \sum_1^{a^k}{}_i \, P_D({}_D\omega_{ij}) = \sum_1^{a^k}{}_i \int_{{}_A\omega_i} {}_\xi\Pi_B \, ({}_B\omega_j) \, P_A \, (d\xi) = \qquad (4.29)$$

$$= \int_{\bigcup_{i=1}^{a^k} {}_A\omega_i} {}_\xi\Pi_B \, ({}_B\omega_j) \, P_A \, (d\xi) = \int_{S_A} {}_\xi\Pi_B \, ({}_B\omega_j) \, P_A \, (d\xi) \triangleq P_B \, ({}_B\omega_j) \, ,$$

and

$$P_D \, ({}_D\varphi_i) = \sum_1^{b^k}{}_j \, P_D({}_D\omega_{ij}) = \sum_1^{b^k}{}_j \int_{{}_A\omega_i} {}_\xi\Pi_B \, ({}_B\omega_j) \, P_A \, (d\xi) =$$

$$= \int_{{}_A\omega_i} \Big[\sum_1^{b^k}{}_j \, {}_\xi\Pi_B \, ({}_B\omega_j) \Big] P_A \, (d\xi) = \int_{{}_A\omega_i} {}_\xi\Pi_B \Big\{ \bigcup_{j=1}^{b^k} {}_B\omega_j \Big\} P_A \, (d\xi) = \int_{{}_A\omega_i} P_A \, (d\xi) = P_A \, ({}_A\omega_i) \, . \qquad (4.30)$$

Eq.s (4.29) and (4.30) define on the elements given by eq. (4.28), i.e. on two fields isomorphic to Ω_B and Ω_A respectively, two "marginal probabilities"; $P_D \, ({}_D\varphi_i)$ coincides with P_A , while $P_D \, ({}_D\vartheta_j)$ is a mean value of ${}_\xi\Pi_B$ through P_A itself. In this way we have introduced in Ω_B a nonconditional probability measure P_B , along with the conditional probabilities ${}_\xi\Pi_B$.

Now we have got three flows: $\{ S_A , \Omega_A , P_A \}$ (which is the source Q_A) is an "input flow"; $\{ S_B , \Omega_B , P_B \}$ is an "output flow"; finally $\{ S_D , \Omega_D , P_D \}$ is a "channel flow" or " compound flow".

It may be shown [28] that if the source $Q_A = \{ S_A ,$ $\Omega_A , P_A \}$ is stationary, then also the compound flow $\{ S_D ,$ $\Omega_D , P_D \}$ is stationary, in the sense that $P_D (T\, {}_D\omega) = P_D$ $({}_D\omega)$ for every cylinder ${}_D\omega$. Under the same assumptions, also the output flow $\{ S_B , \Omega_B , P_B \}$ is stationary.

4. 5. The "Typical Sequences".

Now we intend to state and prove the Source-Coding Theorem in the simplest case, i.e. in case we are given a Discrete Memoryless Source (D M S), whose definition is contained in the definition of section 4-1. We prefer, however, to give a simpler definition of the D M S, in order to stick to the essential features of the problem.

Let us be given a discrete source working with a finite alphabet $A = \{ z_1 , z_2 , \dots , z_a \}$, i.e. a source emitting sequences ξ of the form (4.1) where each x is now a particular z . Let moreover $\mathcal{G} = \{ p_1 , \dots , p_a \}$ be the probability distribution according to which the z' s are generated. Such a discrete source is said to be "memoryless" if each x takes on a value z according to the $\mathcal{G}$ distribution, regardless of which values the preceding x's have taken on. In other words, if $\xi_{(n)} = z_{i_1} , z_{i_2} , \dots , z_{i_n}$ is a n-length sequence, then

$$P(\xi_{(n)}) \overset{def}{=} Prob\,(\xi_{(n)}) = p_{i_1}\, p_{i_2} \cdots p_{i_n} \quad . \qquad (4.31)$$

In emitting an n-length sequence, a D M S actually performs n independent choices among the z's according to the same distribution G . Of course there are a^n distinct n-length sequences output by the source, each having a probability defined by (4.31). If we assume that $1 \geq p_1 \geq p_2 \geq \dots p_a \geq 0$, then ordering the n-sequences according to their decreasing probability yields $z_1, z_1, \dots z_1$ in the first place and z_a, $z_a, \dots z_a$ in the a^n -th place.

As we have anticipated in section 4-4, it is often advisable to encode the sequences output by the source into sequences of letters drawn from a different alphabet (this is the aim the source-encoder was designed for), say $B = \{ y_1, y_2, \dots y_b \}$. Call "codewords" the sequences of y 's associated to the z- sequences.

It is reasonable to assume that the cost of the transmission is proportional to the time the channel is (or the encoders are) used,so that encoding should provide an average length for the codewords as small as possible. Two possibilities are at hand: either we encode more probable n -sequences into shorter codewords and less probable n -sequences into longer codewords (cf. Part One), or we choose a constant length for the codewords. In the second case obviously, we can hope to reduce the average length of the codewords only by reducing the number of source sequences to which distinct codewords are associated.

Although the different-length encoding techniques seem much more attractive, they present many disadvantages [22], which make the constant-length techniques (or "block-coding" techniques) more suitable. Moreover, by carefully choosing the set of n-sequences to which distinct codewords are to be attributed, an (average) length of the codewords arbitrary close to $\frac{H}{\log b}$ * can be achieved, while letting the probability P_a of erroneous decoding by arbitrarily close to zero, provided the length n is large enough. In particular, if $b = 2$, i.e. if we encode into binary codewords, then under precise conditions the (average) length tends to H. Since $H < \log a$ ** this brings about an actual reduction of the codeword length.

As the preceding analysis has pointed out, two parameters are of relevance in the source-encoding methods: one is the erroneous decoding probability and the other is the average length. If to make P_a go to zero as $n \to \infty$ it were necessary to let all n-sequences have distinct codewords, then no

---

* Here, as usual, H means the entropy of the source.

** Actually we have $H = \log a$ if and only if $p_i = \frac{1}{a}$ for $i = 1, 2, \ldots a$. But in this case we cannot hope to reduce the average length of the codewords by a method whatsoever, since all the n-sequences are equiprobable.

reduction in the length would be possible, and the method should be considered as useless.

Fortunately the structure of a D M S is such that its output sequences can be "suitably represented" by a "relative minority" of sequences (which be shall call the "typical sequences"). To be more precise, among all a^n n-sequences there are 2^{nH} sequences whose global probability is as close to 1 as we wish, provided n is large enough. Then it is clear that providing distinct codewords for the 2^{nH} typical sequences and arbitrary codewords for the other sequences, the probability of decoding error is arbitrarily small for n sufficiently large: on reception of a codeword whatsoever we decode it into the corresponding typical sequence, and with probability arbitrarily close to 1 we shall be right.

On the other hand the 2^{nH} typical sequences are a relatively scarce minority among the $a^n = 2^{n \log a}$ sequences, and actually:

$$\lim_{n\to\infty} \frac{2^{nH}}{a^n} = 0 \tag{4.32}$$

as long as $H < \log a$, which we assume to be the case.

The existence of the set of typical sequences follows from a simple application of the law of large numbers. From (4.31), taking the logarithms we get:

$$-\log \text{Prob}\left(\xi_{(n)}\right) = -\log \prod_{j=i}^{n} p_{ij} = -\sum_{1}^{n}{}_{j} \log p_{ij} \quad , \tag{4.33}$$

or, indicating as usual by $I(x) = -\log p(x)$ the self-information of x, also:

$$I(\xi_{(n)}) = \sum_{1}^{n} {}_j \, I(z_{i_j}) \quad . \tag{4.34}$$

Eq. (4.34) tells us that the self-information of an n-sequence output from our D M S is the sum of n independent, identically distributed random variable, coinciding with the self-informations of its components.

The law of large numbers then ensures us that for any $\delta > 0$ there exists an $\varepsilon(n, \delta) > 0$ such that:

$$G_n \left\{ \left| \frac{I(\xi_{(n)})}{n} - H \right| > \delta \right\} \leq \varepsilon(n, \delta) \; ^{*} \tag{4.35}$$

and

$$\lim_{n \to \infty} \varepsilon(n, \delta) = 0 \quad . \tag{4.36}$$

Let T_n be the set of the typical n-sequences, i.e. define:

$$\xi_{(n)} \in T_n \quad \text{if and only if}$$

$$\left| \frac{I(\xi_{(n)})}{n} - H \right| \leq \delta \; ; \tag{4.37}$$

then in force of (4.35) we have:

$$\text{Prob}(T_n) \geq 1 - \varepsilon(n, \delta) \quad . \tag{4.38}$$

---

* Recall that the entropy H of the source is just the mean value of the self-information $-\log p(z_i)$ of the random variable z_i.

On the other hand ineq. (4.37) can be put into the form:

$$n\left[H-\delta\right] \leq I(\xi_{(n)}) \leq n\left[H+\delta\right],$$

or equivalently:

$$2^{-n[H-\delta]} \geq Prob\,(\xi_{(n)}) \geq 2^{-n[H+\delta]}\qquad, \tag{4.39}$$

and these inequalities hold for any typical sequence.

How many sequences are there in T_n ?

An estimate is obtained as follows (M_t indicates the number of the typical sequences):

$$1 \geq G_\pi(T_n) \geq M_t \cdot \min_{\xi_{(n)}\in T_n} Prob\,(\xi_{(n)}) \geq M_t \cdot 2^{-k[M+\delta]},$$

whence:

$$M_t \leq 2^{k[M+\delta]}\qquad. \tag{4.40}$$

On the other hand, in force of (4.38):

$$1-\varepsilon\,(n,\delta) \leq P_x(T_n) \leq M_t \max_{\xi_{(n)}\in T_n} Prob\,(\xi_{(n)}) \leq M_t \cdot 2^{-k[H-\delta]},$$

whence

$$M_t \geq [1-\varepsilon(n,\delta)]\,2^{k[H-\delta]}\qquad. \tag{4.41}$$

Eqs (4.40) and (4.41) jointly yield:

$$[1-\varepsilon\,(n,\delta)]\,2^{k[H-\delta]} \leq M_t \leq 2^{k[H+\delta]}\qquad. \tag{4.42}$$

Taking into account the arbitrariness of δ and eq. (4.36), from (4.38) and (4.42) we can summarize our results in the following

Theorem. The n-length outputs of a D M S can be divided - for n large enough - into two sets. The sequences of the first set, defined by eq. (4.37) are approximately 2^{nH} and each of them has approximately a probability equal to 2^{-nH}. Therefore the first set has an overall probability approximately equal to 1. The precise meaning of the adverb "approximately" is conveyed by eq.s (4.42), (4.39) and (4.38) respectively.

4. 6. The Source - Coding Theorem.

From what we have said in the preceding section, it is apparent that if the $\sim 2^{nH}$ sequences of T_n are provided with distinct codewords and the sequences of T_n^c (complementary set) with arbitrary codewords, then, since the probability of erroneous decoding P_e coincides with the probability of T_n^c, i.e.:

$$P_e = Prob(T_n^c) , \tag{4.43}$$

in force of (4.38) we have

$$P_e \leq \varepsilon(n, \delta) ,$$

i.e., taking into account (4.36), also

$$\lim_{n \to \infty} P_e = 0 , \tag{4.44}$$

while at the same time (4.32) holds.

In other words, as $n \to \infty$ the sequences having a probability greater than 2^{-nH}, or the sequences having a probability smaller than 2^{-nH} are the majority, but their overall probability is negligible.

Very often a prescribed parameter is the encoding rate R (cf. Part One), defined as follows:

$$K = 2^{nR} , \tag{4.45}$$

being K the number of n-sequences for which distinct codewords are to be provided.

Of course the only values of interest for R are those for which

$$K < a^n ,$$

i.e.

$$\frac{\log K}{n} = R < \log a^{*} , \tag{4.46}$$

otherwise distinct codewords would be provided for all the a^n n-sequences, vanifying any attempt towards making the average codeword length close to $H/\log b$.

Now two cases are possible: either $R > H$ or $R < H$. If $R > H$, then of course $K > 2^{nH}$ and consequently distinct codewords can be associated to the typical

---

* Here the logarithms are taken to the base 2 .

sequences. So the following theorem holds:

Theorem (Shannon's Source - Coding Theorem). If the rate R at which a D M S is being encoded is greater than the entropy H of the source, then the probability P_e of incorrect decoding can be made approach zero as close as we wish, provided only the length n of the sequences output by the source is large enough.

Of course it is highly desirable to have also a converse result, confirming the somewhat intuitive feeling one has that the typical sequences are not only sufficient but also necessary to describe the source suitably. Actually such a result has been proved, and can be stated in the following way:

Theorem (Strong converse of the preceding Theorem): If the rate R at which a D M S is being encoded is less than its entropy H, then P_e cannot be made arbitrarily close to zero; rather P_e tends to 1 as $n \to \infty$.

P r o o f. Since $R = \frac{\log K}{n}$ is less than H, putting $K = b^m$, where m is the (uniform) length of the codewords, we have

$$R = \frac{m}{n} \log b < H \quad . \tag{4.47}$$

If for instance

$$\frac{m}{n} \log b \leq H - 2\delta \quad ,$$

being δ a suitable positive number, the codewords are at most $2^{n[H-2\delta]}$, so that we can provide distinct codewords at most for $2^{n[H-2\delta]}$ typical sequences. But since no typical sequence has a probability greater than $2^{-n[H-\delta]}$, the overall probability of the typical sequences for which we are able to provide codewords is

$$2^{-n[H-\delta]} \cdot 2^{n[H-2\delta]} = 2^{-n\delta} . \qquad (4.48)$$

Of course one could also think of providing distinct codewords for some non-typical sequences, e.g. the most probable sequences, which are not among the typical. But since the overall probability of the non-typical sequences is $\varepsilon(n, \delta)$, in this way we are able to provide codewords for sequences having an overall probability not greater than:

$$\varepsilon(n, \delta) + 2^{-n\delta} , \qquad (4.49)$$

so that:

$$P_e \geq 1 - [\varepsilon(n, \delta) + 2^{-n\delta}] . \qquad (4.50)$$

From (4.50) we conclude that $P_e \to 1$ as $n \to \infty$, and the proof is complete.

From this theorem one grasps the great importance of Shannon's entropy H, which comes out in a very natural way. We only mention that in the channel-coding theory Shannon's entropy plays a role as important as here.

To conclude this notes, we remark that Shannon's theorem, although of decisive importance, is an existence theorem and gives no hint as to the construction of implementable codes for given sources in case $R > H$. Moreover nothing is said about the speed at which P_e goes to zero.

References.

[1] Shannon, C.E.: "A Mathematical Theory of Communication", Bell System Technical Journal, 27, pp.379--423, 623-653, 1948.

[2] Middleton, D.: "An Introduction to Statistical Communication Theory", McGraw-Hill, 1960.

[3] Fano, R.M.: "Transmission of Information", the M.I.T. Press, John Wiley and Sons, 1961.

[4] Brillouin: "Science and Information Theory", Academic Press, 1956.

[5] Brillouin: "Scientific Uncertainty and Information", Academic Press, 1964.

[6] Meyer-Eppler W.: "Grundlagen und Anwendungen der Informationstheorie", Springer, (II ed.), 1969.

[7] Huffman, D.A.: "A Method for the Construction of Minimum Redundancy Codes" Proc. I.R.E., 40, p.1098, 1952.

[8] Freinstein, A.:"Foundations of Information Theory", McGraw--Hill, 1958

[9] Ash, R.: "Information Theory", Interscience Publishers, 1965.

[10] Hardy, G.H. - D.E. Littlewood, - G. Polya: "Inequalities", Cambridge University Press, 1952.

[11] Muroga S.: "On the Capacity of a Discrete Channel",

Journal Phys. Soc. Japan, 8, pp.484-494, 1953.

[12] Khinchin A.: "Mathematical Foundations of Information Theory", Dover, 1957.

[13] Faddeiev D.K.: "Zum Begriff der Entropie eines endlichen Wahrscheinlichkeitsschemas" in Arbeiten zur Informationstheorie, I, VEB, 1967.

[14] Lee P.M.: "On the Axioms of Information Theory", Ann. Math. Stat. 35, pp.415-418.

[15] Rényi A.: " On Measures of Entropy and Information", Proceedings of the 4th Berkeley Symposium on Math. Stat. and Probability, vol.I,pp. 547-561, University of California Press, 1960.

[16] Aczél, J. and Z. Daróczy: "Charakterisierung der Entropien positiver Ordnung und der Shannonschen Entropie", Acta Math. Acad. Sc. Hungar., 14, pp. 95-121, 1963.

[17] Rényi A.: "Statistical Laws of Accumulation of Infor mation", Bull. de l'Institut International de Statistique, 33, 1961.

[18] Rényi A.: "Sur la théorie de la recherche aléatoire", Bull. Amer. Math. Soc., 71, pp. 809-828, 1965.

[19] Onicescu O.: "L'énergie informationnelle", C.R. Acad.

Sc. Paris, 263 , pp. 841-842, 1966.

[20] Onicescu O.: "Seminar held at the Istituto di Meccanica", University of Trieste, 1967.

[21] Longo G.and F.Buttazzoni: "Un'interpretazione dell'intensità dell'energia informazionale", Rend. Acc. Naz. Lincei, 44, pp.282-288, 1968.

[22] Gallager R.G.: "Information Theory and Reliable Communication", J. Wiley & Sons, 1968.

[23] Onicescu O.: "Calcolo delle probabilità e applicazioni", Veschi, 1969.

[24] Shilov G.E. and Gurevich B.L.: "Integral, Measure and Derivative: a Unified Approach", Prentice--Hall, 1966.

[25] Rényi A. and J. Balatoni: "Zum Begriff der Entropie", in Arbeiten zur Informationstheorie I, VEB, 1967.

[26] Rényi A.: "On the Dimension and Entropy of Probability Distributions", Acta Math. Acad. Sci. Hung., 10, pp. 193-215, 1959.

[27] Halmos P.R.: "Measure Theory", Van Nostrand, 1950.

[28] Billingsley P.: "Ergodic Theory and Information", John Wiley and Sons, 1965.

[29] Kappos D.: "Strukturtheorie der Warstheinlichkeitsfelder und -räume", Springer, 1960.

Printed in the United States
By Bookmasters